A Statistical Approach
to Neural Networks
for Pattern Recognition

The Wiley Bicentennial—Knowledge for Generations

Each generation has its unique needs and aspirations. When Charles Wiley first opened his small printing shop in lower Manhattan in 1807, it was a generation of boundless potential searching for an identity. And we were there, helping to define a new American literary tradition. Over half a century later, in the midst of the Second Industrial Revolution, it was a generation focused on building the future. Once again, we were there, supplying the critical scientific, technical, and engineering knowledge that helped frame the world. Throughout the 20th Century, and into the new millennium, nations began to reach out beyond their own borders and a new international community was born. Wiley was there, expanding its operations around the world to enable a global exchange of ideas, opinions, and know-how.

For 200 years, Wiley has been an integral part of each generation's journey, enabling the flow of information and understanding necessary to meet their needs and fulfill their aspirations. Today, bold new technologies are changing the way we live and learn. Wiley will be there, providing you the must-have knowledge you need to imagine new worlds, new possibilities, and new opportunities.

Generations come and go, but you can always count on Wiley to provide you the knowledge you need, when and where you need it!

WILLIAM J. PESCE
PRESIDENT AND CHIEF EXECUTIVE OFFICER

PETER BOOTH WILEY
CHAIRMAN OF THE BOARD

A Statistical Approach to Neural Networks for Pattern Recognition

Robert A. Dunne
Commonwealth Scientific and Industrial Research Organization
Mathematical and Information Sciences
Statistical Bioinformatics-Health
North Ryde, New South Wales, Australia

WILEY-INTERSCIENCE
A John Wiley & Sons, Inc., Publication

Published by John Wiley & Sons, Inc., Hoboken, New Jersey.
Published simultaneously in Canada.

For general information on our other products and services or for technical support, please contact our
Customer Care Department within the United States at (800) 762-2974, outside the United States at
(317) 572-3993 or fax (317) 572-4002.

Wiley also publishes its books in a variety of electronic formats. Some content that appears in print may
not be available in electronic format. For information about Wiley products, visit our web site at
www.wiley.com.

Wiley Bicentennial Logo: Richard J. Pacifico

Library of Congress Cataloging-in-Publication Data is available.

ISBN 978-0-471-74108-4

Printed in the United States of America.

10 9 8 7 6 5 4 3 2 1

To my father,
James Patrick

CONTENTS

NOTATION AND CODE EXAMPLES

g_n	$g_n \in \{1, \ldots, Q\}$, the class label, 9
Υ	matrix of weights in the MLP model, 11
Ω	matrix of weights in the MLP model, 11
$\mathrm{mlp}(x)$	the MLP fitted function, 11
$P.H.Q$	parameters describing an MLP model, 11
T	a target matrix, 11
t	a row of the target matrix T, 11
z^*	the MLP output, 11
ω	ω is an element in Ω, 11
υ	υ is an element in Υ, 11
R	$R = (P+1)H + (H+1)Q$ is the total number of weights, 12
ρ_l	sum of squares penalty function, 12
ρ_c	cross–entropy penalty function, 12
$U_X \Lambda_X V_X^T$	SVD of X, 21
$\mathbb{V}$	variance, 35
$\mathbb{E}$	expectation, 35
D_N	data set of size N, 37

P_e^A the apparent error rate, 54

$\mathbb{D}$ deviance, 58

$\kappa(A)$ condition number, 62

$\xrightarrow{D}$ converges in distribution, 123

IC   influence curve, 124

SC   sensitivity curve, 144

There are a number of illustrative examples given at the end of some chapters. Many of these involve code in the S language, a high-level language for manipulating, analyzing and displaying data. There are two different implementations of S available, the commercial system *S–PLUS* (©Insightful Corporation, Seattle, WA http://www.insightful.com) and the open source R (R Development Core Team (2006), http://www.R-project.org). Invaluable references for the use of either system are Venables and Ripley (2000, 2002).

The two systems are implementations of the same S language (Becker et al., 1988), but are free to differ in features not defined in the language specification. The major differences that a user will see are in the handling of third party libraries and in the interfaces with C and Fortran code (Venables and Ripley (2000) give a summary of the differences).

We will use the R system in the examples, and will make use of the many third party contributed software libraries available through the "Comprehensive R Archive Network" (see http://www.R-project.org). Many of these examples will directly translate over to *S–PLUS* .

A number of the functions described here (and the code for the examples) are available through the website http://www.bioinformatics.csiro.au/sannpr. Many of the functions are packaged into the mlp R library.

PREFACE

This work grew out of two stimuli. One was a series of problems arising in remote sensing (the interpretation of multi-band satellite imagery) and the other was the lack of answers to some questions in the neural network literature.

Typically, in remote sensing, reflected light from the earth's surface is gathered by a sensing device and recorded on a pixel by pixel basis. The first problem is to produce images of the earth's surface. These can be realistic looking images but more likely (as they are more useful) they will be "false-colored" images where the coloring is designed to make some feature more visible. A typical example would be making vegetation covered ground stand out from bare ground.

However, the next level of problems in remote sensing concerns segmenting the image. Remote sensing gives rise to many problems in which it is important to assign a class membership label to the vector of pixel measurements. For example, each pixel in the image on page 90 has been assigned to one of the ground cover classes: "pasture"; "wheat"; "salt bush"; . . . , on the basis of its vector of measured light reflectances.

This was a problem that I worked on with the members of the Remote Sensing and Monitoring[1] project in Perth, Australia for many years during the 1990s as a PhD student, visiting academic and group member. The problems tackled by this group were (and are) of great practical importance. The group was heavily involved in developing methods of monitoring and predicting the spread of salinity in the farming areas of the south western corner of Australia. Due to changes in land use (clearing and broadacre farming) many areas of the world are experiencing similar problems with spreading salinity.

[1]http://www.cmis.csiro.au/RSM

To monitor salinity the group utilized the resources of the archived data collected by the LandSat series of satellites (jointly managed by NASA and the U.S. Geological Survey) with information going back to the early 1970s. The approach of the Remote Sensing and Monitoring group was unique in that, due to the large area it monitors and the need to gain an historic perspective on the salinity process, it pursued the use of lower resolution and more limited bandwidth data. This is different to many methodology groups in remote sensing who are keen to explore data from the latest high resolution multi-band sensors. It meant that a major focus of the group was on quite difficult classification problems.

It was in this context that I started looking at neural networks. While I found numerous explanations of how "multi-layered perceptrons" (MLPs) worked, time and time again I found, after working through some aspect of MLPs: implementing it; thinking about it; that it was a statistical concept under a different terminology.

This book then is partly the result of bringing the MLP within the area of statistics. However, not all statisticians are the same. There is a decided emphasis on robustness in this book and very little discussion of Bayesian fitting methods. We are all shaped by the times we lived through and by the questions we have struggled with. I have always worked with data sets that are large either in terms of observations (remote sensing) or variables (bioinformatics). It seems to me that training times for MLP models are long enough already without introducing Bayesian procedures. However there will always be a need to ensure that the model is resistant to anomalies in the data.

How to read this book

Chapters 1–5 describe the MLP model and show how it relates to some other statistical models used for classification tasks, such as linear discriminant analysis and logistic regression. This could form the basis for a self contained graduate level course on MLP models.

Chapters 6 and 7 describe adaptions of the MLP model to situations with large numbers of classes and to some image problems.

chapter	title
1	The Perceptron
2	The Multi–Layer Perceptron Model
3	Linear Discriminant Analysis
4	Activation and Penalty Functions
5	Model Fitting and Evaluation
6	The Task–Based MLP
7	Incorporating Spatial Information into an MLP Classifier

Chapters 8 and 9 investigate the robustness of the MLP model. The reader who is not interested in all the detail of Chapters 8 and 9 could read the summaries on pages 139 and 157 respectively.

Chapters 10 to 13 describe extensions and modifications to the MLP model. Chapter 10 describes a fitting procedure for making the MLP model more robust. Chapter 11 describes a modification for dealing with spectral data. Chapter 12 and 13 further investigate the modification of the weights during the fitting procedure and suggest some extensions to the MLP model based on this.

The book should give a largely self contained treatment of these topics but relies on at least an undergraduate knowledge of statistics and mathematics.

chapter	title
8	Influence Curves for the Multi–layer Perceptron Classifier
9	The Sensitivity Curves of The MLP Classifier
10	A Robust Fitting Procedure for MLP Models
11	Smoothed Weights
12	Translation Invariance
13	Fixed-slope Training

ROB DUNNE

Sydney, Australia

ACKNOWLEDGMENTS

This work owes a lot to Dr Norm Campbell, of the CSIRO Mathematics for Mapping & Monitoring group. Many of the approaches and ideas herein arose from discussions with Norm over the years.

Other people I would like to thank are Dr Harri Kiiveri of the CSIRO, for many helpful discussions and many insights and Professor Ian James, of Murdoch University. I would like also to thank Dr Ron McKay (formerly Vice-Chancellor of the Northern Territory University). Their openness to new ideas and ability to provide insightful criticism were what made this work possible.

The Martin's Farm data set (considered in Section 6.4 and other places) was made available by the Australian Centre for Remote Sensing (ACRES), a division of Geoscience Australia http://www.ga.gov.au.

Roger Clark of the United States Geological Survey also provided data and analysis suggestions.

R. A. D.

CHAPTER 1

INTRODUCTION

Neural networks were originally motivated by an interest in modelling the organic brain (McCulloch and Pitts, 1943; Hebb, 1949). They consist of independent processors that return a very simple function of their total input. In turn, their outputs form the inputs to other processing units. "Connectionism" was an early and revealing name for this work as the capabilities of the brain were felt to lie in the connections of neurons rather than in the capabilities of the individual neurons. Despite many debates over the years about the biological plausibility of various models, this is still the prevailing paradigm in neuro-science.

Important sources for the history of "connectionism" are McCulloch and Pitts (1943), Hebb (1949), Rosenblatt (1962), Minsky and Papert (1969), and Rumelhart et al. (1986). Anderson and Rosenfeld (1988) reproduce many of the historic papers in one volume and Widrow and Lehr (1990) give a history of the development.

However, the modern area of neural networks has fragmented somewhat and there is no attempt at biological plausibility in the artificial neural networks that are used for such tasks as grading olive oil (Goodacre et al., 1992), interpreting sonar signals (Gorman and Sejnowski, 1988) or inferring surface temperatures and water vapor content from remotely sensed data (Aires et al., 2004).

This book, and artificial neural networks in general, sit somewhere in a shared space between the disciplines of Statistics and Machine Learning (ML), which is in turn a cognate discipline of Artificial Intelligence. Table 1.1 summarizes some of the correspondences between the discipline concerns of Machine Learning and Statistics.

A Statistical Approach to Neural Networks for Pattern Recognition by Robert A. Dunne
Copyright © 2007 John Wiley & Sons, Inc.

Table 1.1   Some correspondences between the discipline concerns of Machine
Learning and Statistics.

machine learning		statistics
supervised learning	"learning with a teacher"	classification
unsupervised learning	"learning without a teacher"	clustering

Friedman (1991b) carries the teacher analogy further and suggests that a useful distinction between ML and statistics is that statistics takes into account the fact that the "teacher makes mistakes." This is a very accurate and useful comment. A major strand of this work is understanding what happens when "the teacher makes a mistake" and allowing for it (this comes under the heading of "robustness" in statistics).

Clearly one of the differences between ML and statistics is the terminology, which of course is a function of the history of the disciplines. This is exacerbated by the fact that in some cases statistics has its own terminology that differs from the one standard in mathematics. Another easily spotted difference is that ML has better names for its activities[1].

More substantial differences are that:

- machine learning tends to have a emphasis on simple, fast heuristics. It has this aspect in common with data mining and artificial intelligence;

- following on from the first point, whereas statistics tends to start with a model for the data, often there is no real data model (or only a trivial one) in machine learning.

Breiman in his article "Statistical modelling: The two cultures" (2001) talks about the divergence in practice between Statistics and Machine Learning and their quite different philosophies. Statistics is a "data modeling culture," where a function $f : x \rightarrow y$ is modeled in the presence of noise. Both linear and generalized linear models fall within this culture. However, Machine Learning is termed by Breiman an "algorithmic modeling culture." Within this culture, the function f is considered both unknown and unknowable. The aim is simply to predict a y value from a given x value.

Breiman argues that in recent years the most exciting developments have come from the ML community rather than the statistical community. Among these developments one could include:

- decision trees (Morgan and Sonquist, 1963);

- neural networks, in particular perceptron and multi-layer perceptron (MLP) models;

- support vector machines (and statistical learning theory) (Vapnik, 1995; Burges, 1998);

- boosting (Freund and Schapire, 1996).

---

[1] I believe that this comment was made in the lecture for which Friedman (1991b) are the notes.

Breiman in what, from the discussion[2], appears to have been received as quite a provocative paper, cautions against ignoring this work and the problem areas that gave rise to it.

While agreeing with Breiman about the significance of working with the algorithmic modeling culture we would make one point in favor of the statistical discipline. While these methodologies have been developed within the machine learning community, the contribution of the statistical community to a full understanding of these methodologies has been paramount. For example:

- decision trees were put on a firm foundation by Breiman et al. (1984). They clarified the desirability of growing a large tree and then pruning it as well as a number of other questions concerning the splitting criteria;

- neural networks were largely clarified by Cheng and Titterington (1994); Krzanowski and Marriott (1994); Bishop (1995a) and Ripley (1996) amongst others. The exaggerated claims made for MLP models prior to the intervention of statisticians no longer appear in the literature;

- boosting was demystified by Friedman et al. (1998) and Friedman (1999, 2000);

- support vector machines have yet to be widely investigated by statisticians, although Breiman (2001) and Hastie et al. (2001) have done a lot to explain the workings of the algorithm.

Brad Efron, in discussing Breiman's paper, suggests that it appears to be an argument for "black boxes with lots of knobs to twiddle." This is a common statistical criticism of ML. It arises, not so much from the number of knobs on the black box, which is often comparable to the number of knobs in a statistical model[3], but from the lack of a data model.

When we have a data model, it gives confidence in the statistical work. The data model arises from an understanding of the process generating the data and in turn assures us that we have done the job when the model is fitted to the data. There are then generally some diagnostic procedures that can be applied, such as examining the residuals. The expectation is that, as the data model matches the process generating the data, the residuals left over after fitting the model will be random with an appropriate distribution. Should these prove to have a pattern, then the modelling exercise may be said to have failed (or to be as good as we can do). In the absence of a data model, there seems little to prevent the process being reduced to an ad-hock empiricism with no termination ctiteria.

However the ML community is frequently working in areas where no plausible data model suggests itself, due to our lack of knowledge of the generating mechanisms.

---

[2] As Breiman (2001) was the leading paper in that issue of *Statistical Science*, it was published with comments from several eminent statisticians and a rejoinder from Leo Breiman.

[3] In some instances the number of "knobs" may be fewer for ML algorithms than for comparable statistical models. See Breiman's rejoinder to the discussion.

1.1   THE PERCEPTRON

We will start with Rosenblatt's perceptron learning algorithm as the foundation of the area of neural networks. Consider a set of data as shown in Figure 1.1. This is a 2-dimensional data set in that 2 variables, x_1 and x_2, have been measured for each observation. Each observation is a member of one of two mutually exclusive classes labelled "×" and "+" in the figure.

To apply the perceptron algorithm we need to have a numeric code for each of the classes. We use

$$y = \begin{cases} 1 & \text{for class } \times \text{ and} \\ -1 & \text{for class } + . \end{cases}$$

The model then consists of a function f, called an "activation function," such that:

$$f = \begin{cases} 1 & \omega_0 + \omega^T x \geq 0 \\ -1 & \text{otherwise.} \end{cases}$$

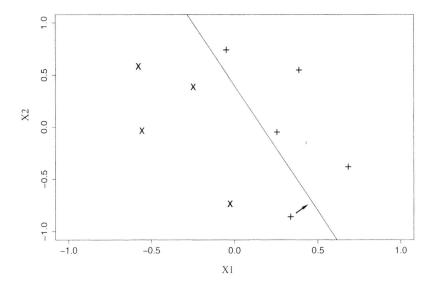

Figure 1.1   *A 2-dimensional data set consisting of points in two classes, labelled "×" and "+". A perceptron decision boundary $\omega_0 + \omega_1 x_1 + \omega_2 x_2$ is also shown. One point is misclassified, that is, it is on the wrong side of the decision boundary. The margin of its misclassification is indicated by an arrow. On the next iteration of the perceptron fitting algorithm (1.1) the decision boundary will move to correct the classification of that point. Whether it changes the classification in one iteration depends on the value of η, the step size parameter.*

The perceptron learning algorithm tries to minimize the distance of a misclassified point to the straight line $\omega_0 + \omega_1 x_1 + \omega_2 x_2$. This line forms the "decision boundary" in that points are classified as belonging to class "×" or "+" depending

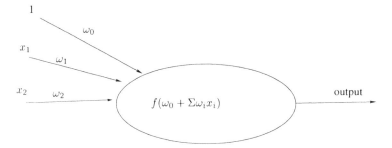

Figure 1.2   The perceptron learning model can be represented as a processing node that receives a number of inputs, forms their weighted sum, and gives an output that is a function of this sum.

on which side of the line they are on. The misclassification rate is the proportion of observations that are misclassified, so in Figure 1.1 it is $1/9$. Two classes that can be separated with 0 misclassification error are termed "linearly separable."

The algorithm uses a cyclic procedure to adjust the estimates of the ω parameters. Each point x is visited in turn and the ωs are updated by

$$\omega_i \leftarrow \omega_i + \eta[y - f(\omega_0 + \omega^T x)]x. \tag{1.1}$$

This means that only incorrectly classified points move the decision boundary. The η term has to be set in advance and determines the step size. This is generally set to a small value in order to try to prevent overshooting the mark.

Where the classes are linearly separable it can be shown that the algorithm converges to a separating hyperplane in a finite number of steps. Where the data are not linearly separable, the algorithm will not converge and will eventually cycle through the same values. If the period of the cycle is large this may be hard to detect.

Where then is the connection with brains and neurons? It lies in the fact that the algorithm can be represented in the form shown in Figure 1.2 where a processing node (the "neuron") receives a number of weighted inputs, forms their sum, and gives an output that is a function of this sum.

Interest in the perceptron as a computational model flagged when Minsky and Papert (1969) showed that it was not capable of learning some simple functions. Consider two logical variables A and B that take values in the set {TRUE, FALSE}. Now consider the truth values of the logical functions AND, OR, and XOR (eXclusive OR, which is true if and only if one of its arguments is true) as shown in Table 1.2.

We can recast the problem of learning a logical function as a geometric problem by encoding {TRUE, FALSE} as {1, 0}. Now for the XOR function, in order to get a 0 classification error the perceptron would have to put the points {1, 1} and {0, 0} on one side of a line and {1, 0} and {0, 1} on the other. Clearly this is not possible (see Figure 1.3).

We note here the very different flavor of this work to traditional statistics. Linear discriminant analysis (see Chapter 3, p. 19, and references therein) is the classical statistical technique for classification. It can not achieve a zero error on the geo-

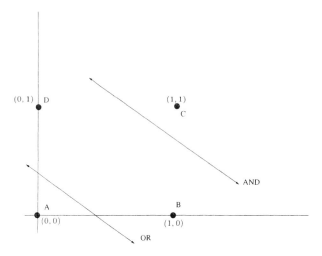

Figure 1.3   The problem of learning a logical function can be recast as a geometric problem by encoding {TRUE, FALSE} as {1, 0}. The figure shows decision boundaries that implement the function OR and AND. The XOR function would have to have points A and C on one side of the line and B and D on the other. It is clear that no single line can achieve that, although a set of lines defining a region or a non-linear boundary can achieve it.

metric XOR problem any more than the perceptron can. However, as far as I know, no statistician has ever shown a lot of concern about this fact.

Table 1.2   Consider two logical variables A and B that can take values in the set {TRUE, FALSE}. The truth values of the logical functions AND, OR, and XOR are shown.

A	B	AND	OR	XOR
T	T	T	T	F
F	F	F	F	F
T	F	F	T	T
F	T	F	T	T

Using a layered structure of perceptron as shown in Figure 2.1 (p. 10) overcame this problem and lead to a resurgence in interest in this research area. These are the "multi-layer perceptrons" (MLPs) that are the topic of this work. They required a different learning algorithm to the single perceptron and require that f be a differentiable function. It was the development of such algorithms that was the first step in their use. This has appeared several times in the literature, common early references being Werbos (1974) and Rumelhart et al. (1986).

Already a large number of questions are apparent: such as:

- what if there are more than two classes;

- what if the classes are not linearly separable – but there is a non-linear decision boundary that could separate them;

- do we want a classifier that performs well on this data set or on a new, as yet unseen, data set? Will they be the same thing?

These questions will be consider in the ensuing chapters.

David Hand has recently written a interesting paper entitled "Classifier Technology and the Illusion of Progress" (Hand, 2006). The progress that he is questioning is the progress of sophisticated techniques like MLPs, support vector machines (Vapnik, 1995; Burges, 1998), and others. He shows that for many examples, the decrease in classification error using sophisticated techniques is only marginal. It may be so small, in fact, that it may be wiped out by the vagaries and difficulties in attendance with real word data sets.

I think that David Hand's caution should be taken to heart. Sophisticated techniques should be used with caution and with an appreciation of their limitations and idiosyncrasies. If there is a good data model available, for example, an understanding that the data are Gaussian, then there may be no justification for using an MLP model.

MLP models have not always been used with an appreciation of their characteristics. The fact that MLPs can be used in a "black box" fashion, and seem to produce reasonable results without a lot of effort being put into modeling the problem, has often led to them being used in this way. It appears that MLPs were being used on hard problems, such as speech recognition and vision, long before any real groundwork was done on understanding the behavior of the MLP as a classifier[4].

This has led to debate in the literature on such elementary points as the capabilities of MLPs with one hidden layer[5], and a lack of understanding of the possible roles of hidden layer units in forming separating boundaries between classes. However, such understanding can be readily arrived at by considering the behavior of the MLP in simple settings that are amenable both to analytic and graphical procedures. In this book the simplest case of two classes and two variables is often used as an example and some points that have been debated in the literature may be amongst the first things that an investigator will notice when confronted with a graphical representation of the output function of an MLP in this simple setting.

The aim of this book is to reach a fuller understanding of the MLP model and extend it in a number of desirable ways. There are many introductions and surveys of multi-layer perceptrons in the literature (see below for references); however, none should be necessary in order to understand this book, which should contain the necessary introduction. Other works that could usefully be consulted to gain insight into the MLP model include Cheng and Titterington (1994), Krzanowski and Marriott (1994), Bishop (1995a), Ripley (1996) and Haykin (1999).

We use a number of examples from the area of remote sensing to illustrate various approaches. Richards and Jia (2006) is a good introduction to this problem area while Wilson (1992) and Kiiveri and Caccetta (1996) discuss some of the statistical issues involved. Once again, this work should be entirely self contained – with as much of the problem area introduced in each example as is needed for a full appreciation of the example. Multi-layer perceptrons have been used in the analysis of remotely sensed data in Bischof et al. (1992), Benediktsson et al. (1995)

---

[4]Early exceptions to this tendency to use MLPs without investigating their behavior are Gibson and Cowan (1990), Lee and Lippmann (1990) and Lui (1990). The situation has been changing markedly in recent years and many of the lacunae in the literature are now being filled.
[5]That is, are they capable of forming disjoint decision regions; see Lippmann (1987).

and Wilkinson et al. (1995). Paola and Schowergerdt (1995) give a review of the application of MLP models to remotely sensed data.

CHAPTER 2

THE MULTI–LAYER PERCEPTRON MODEL

2.1 THE MULTI–LAYER PERCEPTRON (MLP)

We introduce the notation to be used in the following chapters and describe the MLP model, illustrated in Figure 2.1. We end with a brief discussion of how a function approximation model like an MLP can be used for classification.

Say we have a set of training data, $D = \{x_n, t_n\}_{n=1}^{N}$, where each $x_n \in R^P$ is a feature vector of length P, and $g_n \in \{1, \ldots, Q\}$ is the corresponding class label. We form a data matrix X containing the training data

$$
X = \begin{pmatrix} x_1^T \\ \cdot \\ \cdot \\ \cdot \\ x_N^T \end{pmatrix}_{N \times (P+1)}
$$

where $x_n^T \leftarrow (1, x_{n1}, \ldots, x_{nP})$. Hence the data matrix consists of N observations, each of length $P+1$, and each containing a feature vector augmented by the addition of a 1 in the first coordinate position. The 1 is multiplied by a parameter known as the "bias," which is equivalent to the "intercept" or β_0 term in a linear regression.

Comparing Figure 2.1 and Figure 1.2 (p. 5), we can see that the essential structure of the MLP is that of perceptrons interconnected in layers, with the outputs from one layer forming the inputs to the next layer.

A Statistical Approach to Neural Networks for Pattern Recognition by Robert A. Dunne
Copyright © 2007 John Wiley & Sons, Inc.

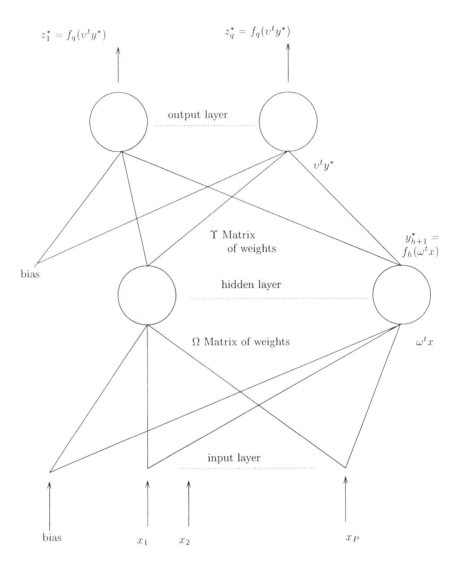

Figure 2.1   In the multi–layer perceptron model, each operational unit, represented in the figure by a circle with a number of input lines, is a perceptron. The outputs of the perceptrons on one layer form the inputs to the perceptrons on the next layer. Each component of the input pattern is presented separately at the nodes of the input layer; the components are weighted, summed, and passed through the perceptron activation function. The outputs of the first, or hidden, layer form the inputs to the nodes of the output layer. While the figure has only two layers of perceptrons, it is of course possible to have an arbitrary number of such layers.

The first stage in constructing the model is to form the product which relates the input layer to the hidden layer,

$$\underset{H \times 1}{y} = \underset{H \times (P+1)}{\Omega} \underset{(P+1) \times 1}{x}$$

where Ω is the first layer of adjustable parameters or weights in the MLP model. Then the function f_H (termed the activation function of the network) is applied to each element of y. In the diagrammatic representation of the MLP (Figure 2.1), this is applied separately at each unit in the hidden and output layers (the circles represent the computational units where this is done).

However, in the algebraic representation, we consider f_H as a mapping from R^H to $(0,1)^H$ that applies the same one–variable function, f_1, to each of its coordinates. The resulting vector $f_H(y)$ is then augmented by the addition of a 1 so that

$$y^* = [1, \{f_H(y_1, \ldots, y_H)\}^T]^T$$

(just as the original data vector x was augmented). This now forms the data input to the next and, in this case final, layer of the network. So we have

$$\underset{Q \times 1}{z} = \underset{Q \times (H+1)}{\Upsilon} \underset{(H+1) \times 1}{y^*}$$

and

$$\mathrm{mlp}(x) = z^* = f_Q(z) = f_Q(\Upsilon[1, \{f_H(\Omega x)\}^T]^T). \tag{2.1}$$

The MLP described here has two layers of adjustable weights and no skip or intra–layer connections and can be described by the parameters $P.H.Q$; that is, P input units, H units in the hidden layer and Q output units. We refer to this as a two–layer MLP as it has two layers of adjustable weights; however, the terminology is not standardized and some authors would call it a three–layer MLP (input, hidden and output) and some a one–layer MLP (one hidden layer). We refer to the weights going from the input to a hidden–layer unit, for example, as the weights "fanning into" the unit and the weights connecting the unit to the output units as the weights "fanning out" of the unit.

While other choices are possible, f_H here is chosen to be the logistic activation function[1],

$$f_H(x) = \frac{1}{1 + \exp(-x)}. \tag{2.2}$$

For general regression problems f_Q may be chosen as the identity function so that $z^* = z$ and the output of the MLP is not restricted to the interval $(0,1)$. As we restrict our attention to classification problems, f_Q will be either the logistic activation function like f_H, or the softmax activation function,

$$f_Q(z_q) = z_q^* = \frac{\exp(z_q)}{\sum_{q_1=1}^{Q} \exp(z_{q_1})}. \tag{2.3}$$

We also form a target matrix T of dimension $N \times Q$, consisting of a high value in position (n, q) if observation n is in class q, and a low value otherwise. When

---

[1]The practice in the neural network literature is to call this the "sigmoid" or the "logistic sigmoid." Other sigmoid functions are possible such as the tanh function.

the high and low values are 1 and 0, respectively, this is referred to as a *one of Q* encoding. The target vector, t_n, is the nth row of T and is an encoding of the class label, g_n, as a target vector of length Q, and t_{nq} is the $(n, q)^{th}$ element of the target matrix.

We index the units on each layer (input, hidden and output) with the indices p, h and q respectively, so that ω_{ph} is the weight between the pth input and the hth hidden–layer unit. When additional indices are required we use p_1, h_1, q_1 etc.

Having set up the MLP model with output z^*, we then choose the weight matrices Υ and Ω to minimize the value of some penalty function. We consider two penalty functions here: one is the sum of squares penalty function

$$\rho_l = \sum_{n=1}^{N} \sum_{q=1}^{Q} \frac{1}{2} (t_{nq} - z_{nq}^*)^2 \tag{2.4}$$

(the factor of $1/2$ is simply for convenience and will be canceled out when derivatives are taken): and the other is the cross–entropy penalty function

$$\rho_c = \sum_{n=1}^{N} \sum_{q=1}^{Q} t_{nq} \log \left(\frac{t_{nq}}{z_{nq}^*} \right).$$

The properties of the penalty and activation functions are discussed in Chapter 4 (p. 35).

In the standard implementation, the weights are initially assigned random values selected uniformly in some small interval, often $(-1, 1)$. While we generally write $\omega \in \Omega$ and $\upsilon \in \Upsilon$, in some contexts, such as in the discussion of function minimization in Appendix A, it is natural to consider ρ as a function of a vector of parameters. In this case we write $\omega = \{\text{vec}(\Upsilon^T)^T, \text{vec}(\Omega^T)^T\}^T$ and index ω by $r = 1, \ldots, R$ where $R = (P + 1)H + (H + 1)Q$ is the total number of weights.

Considering the potential size and complexity of both the data sets and the model, and the fact that a conventional serial CPU is generally used to fit the model, it is clearly of some practical importance that the function minimization be done in an efficient manner. This is perhaps a good point to remark that while it is of practical importance that the function minimization be done in an efficient manner, it is of no theoretical importance. That is, the function being fitted and the method of fitting are quite distinct. This distinction was sometimes lost in the MLP literature where many special purpose function minimization were proposed, and the properties of different network models coupled with different minimization techniques was investigated. The practice now is to simply to adopt an efficient general purpose function minimization scheme from the numerical analysis literature (see comments in Section A.13, p. 256).

2.2 THE FIRST AND SECOND DERIVATIVES

This section can be omitted, unless you intend to implement an MLP model. Fitting the MLP model involves a process of function minimization and many of the function minimization schemes that we consider in Appendix A (p. 245) require the function derivatives. We calculate the first and second derivatives assuming that $\rho = \rho_l$, and that f_Q is the logistic function. Similar computations can be made for other penalty and activation functions. For a single fixed input vector we have, for $q = 1, \ldots, Q$ and $h = 1, \ldots, H + 1$

$$\frac{\partial \rho}{\partial \upsilon_{hq}} = (z_q^* - t_q)\frac{\partial f(z_q)}{\partial z_q}y_h^*, \qquad (2.5)$$

and for $h = 1, \ldots, H$ and $p = 1, \ldots, P + 1$ we have

$$\frac{\partial \rho}{\partial \omega_{ph}} = \sum_{q=1}^{Q}(z_q^* - t_q)\frac{\partial f(z_q)}{\partial z_q}\upsilon_{\{q,h+1\}}\frac{\partial f(y_h)}{\partial y_h}x_p. \qquad (2.6)$$

Note that the derivative $\partial \Upsilon / \partial y_h^*$ picks out the $(h+1)^{th}$ column of Υ. The reason for this is that the hidden layer of the MLP has H inputs (1 for each unit) but $H + 1$ outputs because of the way the bias term has been implemented as an extra unit with a constant output of 1.

The second derivatives are: for $q_1, q_2 = 1, \ldots, Q$ and $h_1, h_2 = 1, \ldots, H + 1$

$$\frac{\partial^2 \rho}{\partial \upsilon_{h_1 q_1}\partial \upsilon_{h_2 q_2}} = \begin{cases} 0 \text{ if } q_1 \neq q_2 \\ \frac{\partial f(z_q)}{\partial z_q}y_{h_1}^*\frac{\partial f(z_q)}{\partial z_q}y_{h_2}^* + (z_q^* - t_q)\frac{\partial^2 f(z_q)}{\partial z_q^2}y_{h_1}^*y_{h_2}^* \text{ if } q_1 = q_2 = q \end{cases}$$

and for $h_1, h_2 = 1, \ldots, H$ and $p_1, p_2 = 1, \ldots, P + 1$

$$\frac{\partial^2 \rho}{\partial \omega_{p_1 h_1}\partial \omega_{p_2 h_2}} = \sum_{q=1}^{Q}\frac{\partial f(z_q)}{\partial z_q}\upsilon_{qh_2+1}\frac{\partial f(y_{h_2})}{\partial y_{h_2}}x_{p_2}\frac{\partial f(z_q)}{\partial z_q}\upsilon_{qh_1+1}\frac{\partial f(y_{h_1})}{\partial y_{h_1}}x_{p_1}$$

$$+ (z_q^* - t_q)\frac{\partial^2 f(z_q)}{\partial z_q^2}\upsilon_{qh_2+1}\frac{\partial f(y_{h_2})}{\partial y_{h_2}}x_{p_2}\upsilon_{qh_1+1}\frac{\partial f(y_{h_1})}{\partial y_{h_1}}x_{p_1}$$

and in the case where $h_1 = h_2 = h$, there is the additional term

$$(z_q^* - t_q)\frac{\partial f(z_q)}{\partial z_q}\upsilon_{qh+1}\frac{\partial^2 f(y_h)}{\partial y_h^2}x_{p_2}x_{p_1}.$$

Then for $p = 1, \ldots, P + 1$ and $q = 1, \ldots, Q$ and for $h_1 = 1, \ldots, H + 1$ and $h_2 = 1, \ldots, H$, we have

$$\frac{\partial^2 \rho}{\partial \upsilon_{h_1 q}\partial \omega_{ph_2}} = \frac{\partial}{\partial \omega_{ph_2}}\left\{(z_q^* - t_q)\frac{\partial f(z_q)}{\partial z_q}y_{h_1}^*\right\},$$

$$= \frac{\partial(z_q^* - t_q)}{\partial \omega_{ph_2}}\frac{\partial f(z_q)}{\partial z_q}y_{h_1}^* + (z_q^* - t_q)\frac{\partial}{\partial \omega_{ph_2}}\left\{\frac{\partial f(z_q)}{\partial z_q}\right\}y_{h_1}^*.$$

In the case where $h_2 + 1 = h_1$, there is the additional term

$$(z_q^* - t_q)\frac{\partial f(z_q)}{\partial z_q}\frac{\partial y_{h_1}^*}{\partial \omega_{ph_2}}.$$

Thus, as we are using the logistic activation function, we can expand the expression and collect like terms to get

$$
\frac{\partial^2 \rho}{\partial \upsilon_{h_1 q} \partial \omega_{p h_2}} = \frac{\partial f(z_q)}{\partial z_q} \upsilon_{q h_2 + 1} \frac{\partial f(y_{h_2})}{\partial y_{h_2}} x_p \frac{\partial f(z_q)}{\partial z_q} y_{h_1}^* +
$$
$$
(z_q^* - t_q) \left(\frac{\partial f(z_q)}{\partial z_q} \upsilon_{q h_2 + 1} \frac{\partial f(y_{h_2})}{\partial y_{h_2}} x_p - 2 f(z_q) \frac{\partial f(z_q)}{\partial z_q} \upsilon_{q h_2 + 1} \frac{\partial f(y_{h_2})}{\partial y_{h_2}} x_p \right) y_{h_1}^*,
$$

plus the additional term

$$
(z_q^* - t_q) \frac{\partial f(z_q)}{\partial z_q} \frac{\partial f(y_{h_1})}{\partial y_{h_1}} x_p,
$$

when $h_2 + 1 = h_1$.

2.3 ADDITIONAL HIDDEN LAYERS

The derivatives for an MLP with more than two layers can be readily calculated using recursive formulae. We investigate this for a three–layer MLP, the architecture of which is described by the parameters $P.H_1.H_2.Q$, in an extension of the previous notation. The units on each layer are indexed with $p, h_1, h_2,$ and q, respectively, and as there is now a new matrix of weights, the weight matrices are renamed as

$$
\underset{H_1 \times (P+1)}{\Omega_1} \qquad \underset{H_2 \times (H_1+1)}{\Omega_2} \quad \text{and} \quad \underset{Q \times (H_2+1)}{\Upsilon}.
$$

The fitted model is then

$$
\mathrm{mlp}(x) = f_Q(\Upsilon\{1, f_{H_2}(\Omega_2[1, \{f_{H_1}(\Omega_1 x)\}^\tau]^\tau)^\tau\}^\tau).
$$

The derivatives can be readily calculated as

$$
\begin{aligned}
\frac{\partial \rho}{\partial \upsilon_{h_2, q}} &= (z_q^* - t_q) \frac{\partial f(z_q)}{\partial z_q} y_{2_{h_2}}^* \\
&= \delta_q y_{2_{h_2}}^* \text{ say,} \\
\frac{\partial \rho}{\partial \omega_{2_{h_1, h_2}}} &= \sum_{q=1}^{Q} \delta_q \upsilon_{\{q, h_2+1\}} \frac{\partial f(y_{2_{h_2}})}{\partial y_{2_{h_2}}} y_{1_{h_1}}^* \\
&= \delta_{h_2} y_{1_{h_1}}^*, \text{ and} \\
\frac{\partial \rho}{\partial \omega_{1_{p, h_1}}} &= \sum_{h_2=1}^{H_2} \delta_{h_2} \omega_{2_{\{h_2, h_1+1\}}} \frac{\partial f(y_{1_{h_1}})}{\partial y_{1_{h_1}}} x_p
\end{aligned}
$$

where, as previously, ω_1, ω_2 and υ are elements of Ω_1, Ω_2 and Υ respectively, and the ys are the inputs to the hidden layers and the y^*s are the augmented outputs from the hidden layers.

2.4   CLASSIFIERS

This section describes how an MLP can be used for classification. It outlines a number of themes that will be considered more closely in the coming chapters.

The map $x \rightarrow z^*$ is a parametric function that maps R^P to R^Q. There are several ways that such a parametric function can be used for classification.

1. We can choose ω to minimize the Euclidean L^2 distance between z^* and t. Then:

    (a) a new observation can be put into the class C_q for which z^* is a maximum. The method is, for a two-class problem, to take x to be a member of group 0 if $mlp(x) > 0.5$ and not if $mlp(x) < 0.5$. We then define the "decision boundary" to be the contour for which $mlp(x) = 0.5$. In the case of a perceptron, the decision boundary will be a point, line or hyperplane such that $\omega^T x = 0$, while in the case of an MLP it may have a more complex structure. As will be shown in Section 4.2 (p. 35), z^* is an unbiased estimate of $P(C_q|x)$ if its functional form is capable of modeling $P(C_q|x)$. If $z^* = \omega^T x$ then the least squares estimate is proportional to Fisher's linear discriminant[2] in the case of 2 groups and gives the best linear approximation to the posterior probabilities of group membership. Efron (1975) showed that in the case of two groups drawn from Gaussian populations, it is more efficient to use linear than logistic regression even though $P(C_q|x)$ is logistic; or

    (b) we can perform a LDA on the $z^*(x_n)$. See Hastie et al. (1994) and Section 3.3, p. 23;

2. We can use the parametric function within a multiple logistic model (softmax) so that

$$P(C_{q_1}|x) = \frac{\exp(z^*_{q_1}(x))}{\sum_{q_2} \exp(z^*_{q_2}(x))}.$$

This involves fitting the model by maximum likelihood (ρ_c in this case, see Chapter 4, p. 35) rather than by least squares; and

3. In the case where the observations are sampled smooth curves, we can form $\arg\min_\omega ||t_n - z^*(\omega, x_n)||_2 = \omega_n$ say, and then

$$\underset{N \times R}{W} = (\omega_1, \omega_2, \ldots \omega_n)^T.$$

We can then do LDA on the parameters W. See Furby et al. (1990) where spline curves are fitted to observed spectral curves and LDA performed on the coefficients of the spline fit. The idea can be extended to other measures of the fitted curve, such as the derivatives, and in this way one can model the shapes of the spectral curves. See Chapter 11 (p. 179) for examples of this.

In addition, we can distinguish a "diagnostic" paradigm for classification where $P(C_q|x)$ is estimated directly. This contrasts with the sampling paradigm (Ripley,

[2]See Chapter 3 for a discussion of linear discriminant analysis (LDA).

1996; Jordan, 1995) where $P(C_q|x)$ may be modeled as

$$P(C_q|x) = \frac{P(x|C_q)\,P(C_q)}{P(x)}$$

using Bayes' rule, and $P(x|C_q)$ is estimated from the data. x is then assigned to the class for which $P(C_q|x)$ is a maximum[3]. It is also possible to distinguish a "discriminatory" paradigm that focuses on learning the decision boundaries between the classes. It is important to be clear which category a classifier belongs to as the assumptions and potential problems are different (Rubenstein and Hastie, 1997).

A ρ_c penalty function with a softmax activation function (case 2) lies within the diagnostic paradigm. However, the common MLP procedure of using multiple independent logistic functions at the output layer is related to case 1(a) and is within the "discriminatory" paradigm, although it will produce (sensible) estimates of $P(C_q|X)$ as the sample size goes to infinity.

2.5 COMPLEMENTS AND EXERCISES

2.1   We fit an MLP model to some data in 2-dimensions, so that we can visualize the decision boundary. We first load the `mlp` library to make the `mlp` function available to us.

```
library(mlp)
```

We then set up the data and plot it.

```
set.seed(123)
data1<-runif(100,-0.5,0.5)
data2<-runif(100,-0.25,0.75)
data3<-cbind(data2,data1)

data4<-runif(50,-1.5,-0.5)
data5<-runif(50,-1,1)
data6<-cbind(data4,data5)

data7<-runif(25,-1.5,1)
data8<-runif(25,1,1.25)
data9<-cbind(data7,data8)

data10<-runif(25,-1.5,1)
data11<-runif(25,-1.25,-1)
data12<-cbind(data10,data11)

data<-rbind(data3,data6,data9,data12)
```

---

[3]This discussion assumes a 0/1 loss function where the cost of a correct classification is 0 and an incorrect classification is 1. It is possible to have a cost matrix specifying the cost of misclassifying a member of C_{q_1} to C_{q_2}. See Titterington et al. (1981) for an example with a cost matrix ascribing differential costs to various outcomes for head-injured patients.

```
x <- seq(-2, 2, length = 20)
y <- seq(-2, 2, length = 20)

target<-make.target(c(100,100),low=0)
#####################################
#set up plot
tt<-c(2,  2, -2,  2,  2, -2, -2, -2)
dim(tt)<-c(2,4)
tt<-t(tt)
plot(tt[,1],tt[,2],type="n",xlab="",ylab="")
points(data[,1],data[,2],pch=15,col=c(rep(2,100),rep(3,100)))
```

The plot is shown in Figure 2.2. We have designed this test example so that we need a non-linear decision region to achieve a zero misclassification error. We fit an MLP of size $P.H.Q = 2.3.2$, that is 2 input nodes, 3 hidden layer nodes and 2 output nodes.

```
set.seed(127)
t1.nn<-mlp(2,3,2,200,data,target)
```

plot.lines plots the lines defined by the rows of Ω. It is only usable when the input space is of 2-dimensional. Similarly, my.mat sets up a matrix of output values for the MLP, suitable for the contour function to plot.

```
plot.lines(t1.nn)
tt<-my.mat(t1.nn)
contour(x,y,t(tt),levels=0.5,add=T)
```

Note that the final decision boundary is a smooth curve. The apparent roughness is caused by the discrete grid used by my.mat, which could be altered (although it will then require more time for the computation).

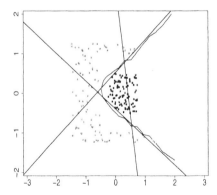

Figure 2.2   The lines defined by the rows of Ω and the 0.5 contour for the MLP outputs.

There are a number of other things we can do to explore MLPs with this data set. You should try

- different starting values (just change the value of set.seed);

- rotate the data and see what happens

    ```
    temp<-matrix(c(cos(pi/4),sin(pi/4),-sin(pi/4),cos(pi/4)),2,2)
    rotated.data<-t(temp %*% t(data))
    ```

- fit the model with more (or less) hidden layer units.

2.2   The Iris data set (Fisher, 1936) is available in the MASS library. The data set consists of 4 measurements (sepal length, sepal width, petal length and petal width) on each of 50 individuals of 3 species: Iris setosa; versicolor; and virginica.

Fit an MLP model to predict membership of the three classes. The help file for nnet (MASS library) gives an example of fitting the model on 1/2 of the data and testing the model fit on the other 1/2. Try this.

CHAPTER 3

LINEAR DISCRIMINANT ANALYSIS

In this chapter we review linear discriminant analysis (LDA), a classical statistical procedure with a number of desirable features. We also look at some flexible extensions to LDA and consider the connection between LDA and the MLP model.

LDA was developed in the case of two classes by Fisher (1936) and extended to multiple classes by Rao (1948). See Rao (1973, pp. 574–580) or Mardia et al. (1979) for a standard exposition. The method is often very successful and we note that in the STATLOG project (Mitchie et al., 1994), a large-scale comparison of classifiers on a wide variety of real and artificial problems, LDA was among the top three classifiers for 11 of the 22 data-sets[1].

We begin by formulating the criterion. Assume that there are Q classes indexed by $q = 1, \ldots, Q$ with N_q observations in the qth class, $\sum_q N_q = N$, and $\underset{N \times Q}{T}$ and $\underset{N \times P}{X}$ are the usual target and data matrices. Assume that the columns of X are mean centered or else set

$$X \leftarrow X - \frac{1}{N} \underset{N \times N}{1} X.$$

---

[1] It is quite complicated to summarize the results of this large set of trials, and there was no outright winner. However, it is fair to say that the two simplest techniques considered, LDA and nearest neighbor, did better than many more sophisticated techniques!

A Statistical Approach to Neural Networks for Pattern Recognition by Robert A. Dunne
Copyright © 2007 John Wiley & Sons, Inc.

Define $P_T = T(T^T T)^{-1} T^T$, the projection matrix onto the column space of T, and then

$$\Sigma_T \underset{P \times P}{=} N^{-1} X^T X,$$
$$\Sigma_B = (Q-1)^{-1} (P_T X)^T (P_T X), \text{ and} \qquad (3.1)$$
$$\Sigma_W = (N-Q)^{-1} ((I-P_T)X)^T ((I-P_T)X),$$

the total covariance matrix, between-classes covariance matrix (the covariance of the projection onto the column space of T) and the within-classes covariance matrix (the covariance of the projection onto the null space of T) respectively. Defining M at the $Q \times P$ matrix of class means, $M = (T^T T)^{-1} T^T X$, we have

$$\Sigma_B = (Q-1)^{-1} (TM)^T (TM), \text{ and}$$
$$\Sigma_W = (N-Q)^{-1} (X-TM)^T (X-TM).$$

Σ_W is assumed to be the same for all the classes and $N\Sigma_T = (Q-1)\Sigma_B + (N-Q)\Sigma_W$.

The task in a LDA is to find A such that the ratio of determinants

$$\frac{|A^T \Sigma_B A|}{|A^T \Sigma_W A|} \qquad (3.2)$$

is maximized (see Problem 3.1, p. 30). This maximizes the between-classes covariance relative to the within-classes covariance in the linear space spanned by the columns of A. The columns of A define the "linear discriminants" also called the "canonical variates". Note that there is nothing in this formulation about classifying observations. The data is projected into a linear space selected according to a criteria. In that space it may be classified or visualized as desired. The way that the linear space has been selected may make the operations easier then in the original space.

Equation 3.2 can be maximized by finding A such that $|A^T \Sigma_B A|$ is maximized subject to $A^T \Sigma_W A = I$. Using a Lagrange multiplier the problem can be written as

$$\Sigma_B A - \lambda \Sigma_W A = 0$$

so that

$$(\Sigma_W^{-1} \Sigma_B - \lambda I) A = 0,$$

assuming that Σ_W is invertible. This is a "generalized eigen–decomposition" problem. A is of dimension $P \times r$ where $r \le \min\{P, Q-1\}$. In the case of two classes, an equivalent procedure is to maximize $\text{trace}(A^T \Sigma_B A)$ subject to $A^T \Sigma_W A = I$.

When the classes have Gaussian distributions with common Σ_W and equal priors, (3.2) can be derived from the Bayesian classification rule discussed in Section 7.1 (p. 93). However, even when these conditions are not true, it is a plausible and appealing criterion.

The usual computational approach is to take the eigen-decomposition of Σ_W

$$\Sigma_W = U_{\Sigma_W} \Lambda_{\Sigma_W} U_{\Sigma_W}^T, \qquad (3.3)$$

where Λ is diagonal and the columns of U are orthogonal, so that

$$\Lambda_{\Sigma_W}^{-1/2} U_{\Sigma_W}^\tau \Sigma_W U_{\Sigma_W} \Lambda_{\Sigma_W}^{-1/2} = I. \tag{3.4}$$

This "spheres" the data, that is the within-class covariance matrix of the data is the identity. We then apply the same transform to Σ_B, namely

$$\Sigma_B^\star = \Lambda_{\Sigma_W}^{-1/2} U_{\Sigma_W}^\tau \Sigma_B U_{\Sigma_W} \Lambda_{\Sigma_W}^{-1/2} \tag{3.5}$$

and carry out an eigen-decomposition of the transformed between-classes covariance matrix $\Sigma_B^\star$, followed by a transformation back to the original variables. Another way of thinking about LDA is that it is a principle component analysis of the between-class covariance matrix after "sphering" the data. Campbell and Atchley (1981) give a geometric interpretation of LDA based on this interpretation.

3.1   AN ALTERNATIVE METHOD

This section can be skipped at at first reading. Some of the equations are used in Section 3.3 (p. 23) to show a relationship between LDA and regression. The reader could refer back to those equations as needed.

We consider the method outlined in Ripley (1996, §3.1), which seems a little more complicated, and obscures the direct geometric interpretation, but makes a connection with a regression based approach that we consider further in Section 3.3 (p. 23). We first rescale the variables to have unit variance by taking the singular value decomposition (SVD) of X (Golub and van Loan, 1982), $X = U_X \Lambda_X V_X^\tau$; we set $S = \sqrt{N} V_X \Lambda_X^{-1}$ and work with the rescaled variables

$$X^* = XS. \tag{3.6}$$

Then $(X^*)^\tau X^* = NI$.

Note that

$$T^\tau T = \begin{pmatrix} N_1 & \cdots & & \cdots \\ \cdots & N_2 & & \cdots\cdots \\ & \cdots\cdots\cdots\cdots\cdots & \\ \cdots\cdots\cdots & & N_Q \end{pmatrix}$$

and set

$$G = \text{diag}\left(\sqrt{\frac{N}{N_1}}, \sqrt{\frac{N}{N_2}}, \ldots, \sqrt{\frac{N}{N_Q}} \right),$$

so that $GT^\tau TG = NI$. Write M^* as the transformed matrix of class means

$$M^* = (T^\tau T)^{-1} T^\tau X^* = N^{-1} G^2 T^\tau X^*. \tag{3.7}$$

Let the SVD of $\tilde{M} = G^{-1} M^*$ be

$$\tilde{M} = U_{\tilde{M}} \Lambda_{\tilde{M}} V_{\tilde{M}}^\tau \tag{3.8}$$

so that

$$\begin{aligned} (Q-1)\Sigma_B^\star &= (TM^*)^\tau (TM^*) \\ &= V_{\tilde{M}} \Lambda_{\tilde{M}}^\tau U_{\tilde{M}}^\tau (GT^\tau TG) U_{\tilde{M}} \Lambda_{\tilde{M}} V_{\tilde{M}}^\tau \\ &= N V_{\tilde{M}} \Lambda_{\tilde{M}}^2 V_{\tilde{M}}^\tau. \end{aligned}$$

Similarly, since the columns of X^* are mean centered,

$$
\begin{aligned}
(N - Q)\Sigma_W^* &= (X^*)^T X^* - (Q - 1)\,\Sigma_B^* \\
&= NI - NV_{\tilde{M}}\Lambda_{\tilde{M}}^2 V_{\tilde{M}}^T \\
&= NV_{\tilde{M}}(1 - \Lambda_{\tilde{M}}^2)V_{\tilde{M}}^T.
\end{aligned}
$$

The original problem (3.2) reduces to finding a linear combination, a, of the rescaled variables X^* that maximizes the ratio

$$
\frac{a^T V_{\tilde{M}}\Lambda_{\tilde{M}}^2 V_{\tilde{M}}^T a}{a^T V_{\tilde{M}}(1 - \Lambda_{\tilde{M}}^2)V_{\tilde{M}}^T a}. \tag{3.9}
$$

Setting $b = V_{\tilde{M}}^T a$, (3.9) becomes

$$
\frac{\sum \lambda_i^2 b_i^2}{\sum (1 - \lambda_i^2) b_i^2},
$$

which is maximized when only b_1 is non-zero as the λ_i are in descending order of magnitude. On the original variables, a is proportional to the first column of $SV_{\tilde{M}}$ and so the first column of $X^{**} = X^* V_{\tilde{M}} = XSV_{\tilde{M}}$ gives the discriminant scores[2] on the first linear discriminant. As in the variables X^{**} both Σ_W^{**} and Σ_B^{**} (the within and between covariance matrices in the variable X^{**}) are diagonal, subsequent columns of X^{**} give further linear combinations which maximize (3.9) subject to being uncorrelated with the previous linear discriminants. This gives a total of r discriminant functions where $r \le \min\{P, Q - 1\}$. The ratios

$$
\frac{(N - Q)\lambda_i^2}{(Q - 1)(1 - \lambda_i^2)}, \tag{3.10}
$$

where λ are the singular values of $\tilde{M}$, measure between- to within-class variances on the ith canonical variate. If these quantities are large for the first few canonical variates, then it may be possible to summarize the data with low-dimensional graphical displays of the class means.

3.2 EXAMPLE

We demonstrate the use of LDA with the following example, originally from Campbell and Mahon (1974), see Problem 3.4, p. 30. Rock crabs assigned to the species *leptograpsus variegatus* have two distinctive color forms, orange and blue. They are an instance of a "polymorphic" genera in which "species present an inordinate amount of variation and hardly two naturalists can agree which forms rank as species and which as varieties" (Charles Darwin quoted by Mahon (1974, p. v)).

The aim of the analysis is to see if a linear combination of morphological variables completely separates the orange and blue crabs. If they can be so separated, the absence of any intermediate forms will be taken as evidence that there is no genetic

---

[2]The "discriminant scores" are just the projections onto the canonical variates. See Problem 3.4 (p. 30), where this is done in R, so that one can examine the outputs at every stage. Many of these names are historical within the statistics community and differ from the usage in mathematics texts.

flow between the two groups; that is, they are not inter-breeding populations and hence are different species. This is a discrimination problem in that the aim is to describe the differences between the classes, not to classify a new example to the classes "blue crab" or "orange crab".

The data set consists of 5 measurements made on each of 200 crabs collected at North Mole, Fremantle, Western Australia. The set of 200 animals was composed of 50 of each of the groups: orange males, blue males, orange females and blue females (abbreviated as OM, BM, OF, and BF). The variables were selected so that they could be measured easily and accurately and are set out, with their abbreviations, in Table 3.1. Mahon (1974) gives details of how the measurements were taken.

Table 3.1   Abbreviations of crab measurements.

FL	front lip
RW	rear width
CL	carapace length
CW	carapace width
DB	body depth

None of the five measurements individually provides any discrimination between the two varieties as each one overlaps throughout most of its range. Similarly, bivariate scatterplots indicate that all variables are highly correlated and fail to completely separate the varieties, although the overlap is not great for some choices of variables (Campbell and Mahon, 1974).

As the data are lengths they were converted to a log scale. Expression (3.10) takes the values $\{5.00 \times 10^2, 2.73 \times 10^2, 6.80, 7.91 \times 10^{-25}\}$ showing that the first two canonical variates account for almost all of the variation. A plot of the data in canonical variate space (Figure 3.1) shows the two forms well separated and the sexes separated to some extent. The absence of intermediate forms can be used to argue that the orange and blue crabs are not inter-breeding populations.

3.3   FLEXIBLE AND PENALIZED LDA

Hastie (1994) identifies two distinct problems with LDA:

1. in the case where there are "lots" of data relative to the dimensionality of the problem, the restriction to linear decision boundaries may be overly restrictive;

2. where there are many correlated features, LDA may use too many parameters and overfit the data, estimating the parameters with high variance[3].

Hastie (1994); Hastie et al. (1994, 1995); Hastie and Tibshirani (1996), and Hastie et al. (1999) have produced a body of work aimed at a unified treatment of these problems. We consider this extension to LDA in some detail because it produces a set of classifiers that address one of the problems that the MLP model addresses,

[3]MLP models also suffer from this problem. One of the extensions to the MLP that is considered in this book (Chapter 11, p. 179) is designed to deal with large numbers of correlated features.

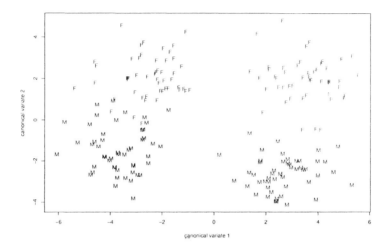

Figure 3.1   The first two canonical variates for the *Leptograpsus* data. The individuals are represented by the codes shown in Table 3.1.

namely flexible decision regions[4]. Flexible and penalized LDA will be used as a comparator for MLP models in later chapters.

If we regress T on the rescaled variables X^* (3.6, p. 21), then $B = (X^{*T}X^*)^{-1}X^{*T}T$. Now

$$T^T X^* = T^T T M^* \quad \text{using (3.7)}$$
$$= T^T T G U_{\hat{M}} \Lambda_{\hat{M}} V_{\hat{M}}^T \quad \text{using (3.8)}$$
$$= N G^{-1} U_{\hat{M}} \Lambda_{\hat{M}} V_{\hat{M}}^T,$$

and therefore

$$B = N^{-1} V_{\hat{M}} \Lambda_{\hat{M}} U_{\hat{M}}^T G^{-1} N$$
$$= V_{\hat{M}} \Lambda_{\hat{M}} U_{\hat{M}}^T G^{-1}.$$

So the predicted values X^*B are of the form $X^* V_{\hat{M}} \Lambda_{\hat{M}} U_{\hat{M}}^T G^{-1} = X S V_M C$ say, where C is a full-rank matrix[5] and so spans the same space as the linear discriminants. However, while they span the same space, the regression coefficients do not give the linear discriminant classifier[6] as the qth column of B is

$$B[,q] = \frac{N_q}{N-1} \Sigma_T^{-1} M^*[q,]$$

[4]Hastie (1994) mentions specifically responding to the challenge of neural networks in the area of flexible decision regions as a motivation in the development of flexible LDA.
[5]$\text{rank}(C) = \text{rank}(\Lambda_{\hat{M}} U_{\hat{M}}^T G^{-1}) = r$ the dimension of the discriminant space.
[6]In the case of $Q = 2$ and target values of $\pm(-\sqrt{N_2/N_1}, \sqrt{N_1/N_2})$, it can be shown that the regression coefficients are proportional to the linear discriminant, and thus give the same line onto which the data are projected. See Ripley (1996, §3.2).

where we note the appearance of Σ_T (recall that M^* is the matrix of group means). Hence the discriminant function that is being minimized is

$$\frac{|A^T \Sigma_B A|}{|A^T \Sigma_T A|}. \tag{3.11}$$

Ripley (1996, §3.2) comments that there are ways to adjust for this (explored in Breiman and Ihaka, 1984) but that the simplest course of action is to apply standard linear discriminant analysis in the space of the fitted values.

So regression on T followed by LDA gives the same results as LDA. Not at first sight a very interesting result. However, the advantage of this formulation is that it suggests some generalizations of LDA by modifying the regression step. For example, if the design matrix X is expanded with the quadratic and cross-product terms of the features, and LDA performed on the fitted values of a regression of T on X, then the linear decision boundaries in the enlarged feature space map down to quadratic boundaries in the original space (see Problem 3.5 at the end of this chapter, p. 32).

This can be seen in Figure 3.2, where we have two zero-centered classes, one with $||x|| < 1.14$ and the other with $||x|| > 1.14$. Extending the feature set with the quadratic and cross–product terms and using LDA gives the elliptical decision boundary shown in the figure. We get an identical results by doing LDA on the fitted values after regressing T on the extended feature matrix. Hence LDA can be extended to a more flexible tool; however, it is necessary to supply the appropriate basis functions in order to do this successfully. Quadratic discriminant analysis (QDA), where the Σ_W are not assumed equal gives a somewhat different ellipse.

Hastie et al. use the term "flexible discriminant analysis" (FDA) for this technique. These methods can be used to generalize LDA and remove the restriction of linear decision boundaries, depending on the regression technique chosen as a building block. Ripley (1994c, §3) gives an example of a projection pursuit regression followed by a robust LDA on the fitted values.

The question arises – why perform an LDA, why not just classify to the maximum row element in $\hat{T}$? Hastie et al. (1999) consider a model where we can add powers of feature variables. For Q classes arranged in a line, we would need an order Q polynomial to separate them. In order to avoid this we can do LDA on the fitted values arising from a regression model with a polynomial of order less than Q.

We note that non-parametric regression techniques such as MARS (Friedman, 1991a), smoothing spline and additive spline models, projection pursuit regression (Friedman and Stuetzle, 1981; Huber, 1985) and MLP models with f_Q as the identity function (and a weight decay penalty term, Chapter 5, p. 53), can be interpreted as expanding the feature set by a basis expansion and doing a (penalized) linear regression in the enlarged space.

FDA then addresses problem (1). To address problem (2) Hastie et al. (1995) suggest that, analogous to the use of a flexible regression model, we use a penalized regression such as

$$T = X\beta + \lambda L(\beta),$$

where $L(\beta)$ is an appropriate penalty term, perhaps one that penalizes large β.

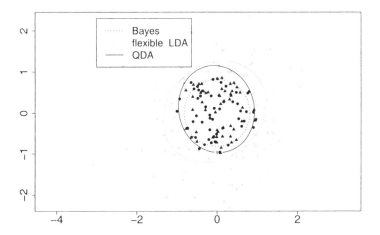

Figure 3.2   Two classes, one with $||x|| < 1.14$ and the other with $||x|| > 1.14$ Extending the feature set with the quadratic and cross–product terms and then performing a LDA gives the elliptical decision boundary shown. The quadratic decision surface is also shown.

3.4   RELATIONSHIP OF MLP MODELS TO LDA

A single hidden layer MLP with f chosen as the identity function and ρ as the sum of squares gives an iterative solution to

$$\arg\min_{\Omega} || \underset{N \times Q}{T} - \underset{N \times P+1}{X} \underset{P+1 \times Q}{\Omega^{\tau}} ||_2^2 \tag{3.12}$$

and produces the mean square classifier which maximizes

$$\frac{|\Omega^{\tau} \Sigma_B \Omega|}{|\Omega^{\tau} \Sigma_T \Omega|}.$$

The classifier gives the best linear approximation to the posterior probabilities and a new observation is put into the class corresponding to the maximum output. We note that for two groups, if the populations are Gaussian, this model is more efficient than logistic regression even though $\mathbb{E}[T|X]$ is logistic not linear (Efron, 1975).

With two hidden layers, provided $H = r = \text{rank}(\Sigma_B)$, and f chosen as the identity function (and ρ as the sum of squares) the MLP model reduces to the same mean square classifier (Fukunaga, 1990)

$$\arg\min_{\Omega, \Upsilon} ||T - X\Omega^{\tau} \Upsilon^{\tau}||_2^2 \tag{3.13}$$

(ignoring the bias terms at the hidden layer), as the product of two full rank matrices is a full rank matrix. See Gallinari et al. (1988) and Baldi and Hornik (1995) for a discussion of the properties of linear MLPs.

Webb and Lowe (1990) extend the result to MLPs with non-linear f_H. Choosing optimum weights to minimize ρ chooses Ω such that the transformation from the data to the hidden layer outputs maximizes trace$(\tilde{\Sigma}_B \tilde{\Sigma}_T^+)$ in the space spanned by the outputs of the hidden layer. However, we still need to know how the non-linear transformations of the MLP work in the process of producing a decision boundary.

3.5   LINEAR CLASSIFIERS

Note that the derivation of the linear discriminant in Equations (3.2) to (3.9) does not depend on any assumptions about the distributions of the classes. If we assume a Gaussian distribution then the LDA can be derived via a maximum likelihood argument. Just considering two classes, with distributions $N(\mu_1, \Sigma_w)$ and $N(\mu_2, \Sigma_w)$, maximizing (3.2) reduces to calculating $(\mu_1 - \mu_2)^T \Sigma_W^{-1}$. We write $(\mu_1 - \mu_2)^T \Sigma_W^{-1} x = g(x)$, the discriminant scores. The likelihood of x being in class 1 versus class 2 is

$$\frac{P_1(x)\pi_1}{P_2(x)\pi_2},$$

where $P_i(x)$ and π_i are the densities and priors for classes $i = 1, 2$. Taking logs and rearranging we have

$$(\mu_1 - \mu_2)^T \Sigma_W^{-1} x - \frac{1}{2}(\mu_1 \Sigma_W^{-1} \mu_1 - \mu_2 \Sigma_W^{-1} \mu_2) - \log(\pi_2) + \log(\pi_1)$$

$= g(x) - c$ say. The classification rule is then:

$$\text{if } g(x) > c \rightarrow x \in \text{class 1};$$
$$\text{if } g(x) < c \rightarrow x \in \text{class 2}.$$

It would appear that a common practice in the literature (see Brown et al., 2000, for an example) is to form $(\hat{\mu}_1 - \hat{\mu}_2)^T (\hat{\Sigma})^{-1}$ and estimate c by evaluating the classification criterion at points along the linear discriminant.

If we do not wish to assume Gaussian distributions with equal covariance matrices we have a number of options. We can use quadratic discriminant analysis or some other more general method which makes no distributional assumptions. Another option is to continue to use LDA. Despite the fact that the underlying assumptions are not met, LDA may still perform well due to the small number of parameters that need to be estimated, as compared to quadratic discriminant analysis. Yet another option is to modify LDA to give the best linear classifier in the case where the covariance matrices are not equal.

Consider two classes, 1 and 2, for which discriminant scores have been calculated and a c value selected. Figure 3.3 shows the probability of a member of class 1 being misclassified as a member of class 2 for the given c value, and Figure 3.4 shows the probability of a member of class 2 being correctly classified. The curve formed by the locus of points $\{P_c(2|1), P_c(2|2)\}$ is called the receiver operating characteristic (ROC) curve and is shown in Figure 3.5. A better classifier will result in a ROC with smaller $P(2|1)$ for the same $P(2|2)$. The measure of classification accuracy that we have generally used $(P_c(2|1) + P_c(1|2))$ is minimized by the point on the

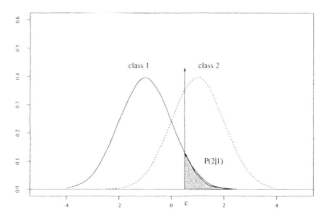

Figure 3.3 For two classes, class 1 and class 2, the probability of a member of class 1 being misclassified as a member of class 2 is shown for a given c value.

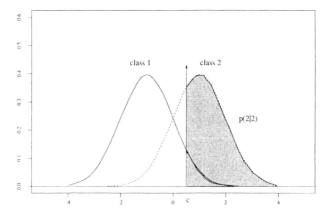

Figure 3.4 For two classes, class 1 and class 2, the probability of a member of class 2 being classified a member of class 2 is shown for a given c value.

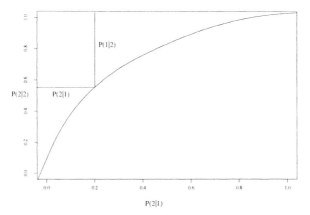

Figure 3.5   The ROC curve for the two classes shown in Figure 3.4. The curve is formed by the locus of points $\{P_c(2|1), P_c(2|2)\}$.

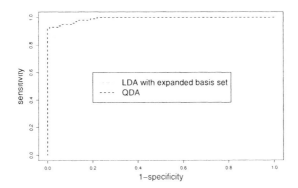

Figure 3.6   The ROC curve for the two classes shown in Figure 3.2, with generalized LDA and QDA.

ROC curve such that length of the sum of the line segments marked $P(2|1)$ and $P(1|2)$ in Figure 3.5 is minimized.

If we can assume Gaussian distributions with proportional covariance matrices then LDA, that is $(\mu_1 - \mu_2)^T (\Sigma_1 + \Sigma_2)^{-1}$, will give the optimum discriminant (Su and Liu, 1993). If we can assume Gaussian distributions with unequal covariance matrices then Anderson and Bahadur (1962) and Clunies-Ross and Riffenburgh (1960) show that the optimum discriminant is given by $(\mu_1 - \mu_2)^T (t_1\Sigma_1 + t_2\Sigma_2)^{-1}$. Anderson and Bahadur show that no linear procedure is superior to all others across the range of the ROC curve and give iterative procedures for determining t_1, t_2 and c in the following special cases:

- minimizing one probability of misclassification for a specified probability of the other;

- minimizing the maximum of the misclassification probabilities;

- minimizing the sum of the misclassification probabilities with given prior probabilities.

This amounts to evaluating the ROC curve at specific points, see also Anderson (1984, §6.10.2). Cooke and Peake (2002) extend the procedure of Anderson and Bahadur to the case where the two classes are of unspecified distribution but their means and covariances are known and also give a simulation method for evaluating the whole ROC curve. Su and Liu (1993) show that the linear discriminant $(\mu_1 - \mu_2)^T (\Sigma_1 + \Sigma_2)^{-1}$ is optimum under the criterion of maximum area under the ROC curve.

3.6   COMPLEMENTS AND EXERCISES

3.1   Consider the 2×2 matrix

$$\begin{bmatrix} 1 & 2 \\ 3 & 4 \end{bmatrix}$$

Write down the determinant of this matrix. Draw the parallelogram spanned by its row vectors, and calculate its area.

For further in information about determinants see Apostol (1967) or other books on linear algebra. The determinant has a "physical" interpretation as the n-dimensional volume of a matrix. However surprisingly, many texts gloss over the interpretation of determinants to concentrate on the calculation and rules for manipulation them. If the linear map $f : \mathbb{R}^n \to \mathbb{R}^m$ is represented by the $m \times n$ matrix A, and M is a measurable subset of $\mathbb{R}^n$ then the n-dimensional volume of $f(M)$ is given by

$$\sqrt{\det(A^T A)} \times \text{volume}(M)$$

Hence (3.2) (p. 20) is a ratio of two volumes, and we are seeking A such that this ratio is maximized.

3.2   Using equations 3.1 to 3.5, calculate Σ_B and Σ_W for the iris data (p. 18) and carry out a LDA.

3.3   Assume that $P(X|C_1) \sim N(\mu_1, \Sigma_1)$ and $P(X|C_2) \sim N(\mu_2, \Sigma_2)$. If we allocate an observation to a group on maximum posterior probability we have the discrimination rule

$$\frac{f_1(x)}{f_2(x)} > k.$$

Substitute the two densities into this formula and derive the quadratic discrimination function. Make the simplifying assumption that $\Sigma_1 = \Sigma_2 = \Sigma$, and simplify the rule to:

$$\text{allocate } x \text{ to } \begin{cases} C_1 & \text{when } (\mu_1 - \mu_2)^T \Sigma^{-1}[x - 1/2(\mu_1 + \mu_2)] > \log k, \\ C_2 & \text{otherwise.} \end{cases}$$

3.4   We consider the Leptograpsus example, discussed in Section 3.2, p. 22. The data set is available in the MASS library (Venables and Ripley, 2002),

```
library(mlp)
library(MASS)
data(crabs)

training.data<-crabs[,4:8]
X<-scale(log(training.data),center=T,scale=F)
X<-as.matrix(X)#
X[1:3,]
          FL          RW          CL         CW         BD
1 -0.6276484 -0.6210677 -0.6637943 -0.625407 -0.6633421
2 -0.5447608 -0.4819549 -0.5467017 -0.534893 -0.6077723
3 -0.5003090 -0.4690515 -0.4981746 -0.460785 -0.5680319

training.labels<-rep(1:4,each=50)

target<-make.target(c(50,50,50,50),low=0)

Q <- dim(target)[2]
N <- dim(X)[1]
P<-dim(X)[2]
means <- solve(t(target) %*% target) %*% t(target) %*% X
m.center<-apply(X,2,mean)
X<-scale(X,center=m.center,scale=F)
Mu <- solve(t(target) %*% target) %*% t(target) %*% X
SigmaB <- (t(target %*% Mu) %*% (target %*% Mu))/(Q - 1)
SigmaW <- (t(X) %*% X - (Q - 1) * SigmaB)/(N - Q)
eigen.SigmaW <- eigen(SigmaW, symmetric = T)
UU <- eigen.SigmaW$vectors
ee <- eigen.SigmaW$values
ee.inv <- 1/ee
Ustar.inv <- diag(sqrt(ee.inv)) %*% t(UU)
Bstar <- Ustar.inv %*% SigmaB %*% t(Ustar.inv)
aa <- eigen(Bstar, symmetric = T)$vectors
cv <- UU %*% diag(sqrt(ee.inv)) %*% aa

#the first two canonical variates are
> cv[,1:2]
           [,1]        [,2]
[1,]   31.217207  -2.851488
[2,]    9.485303 -24.652581
[3,]    9.822169  38.578804
[4,]  -65.950295 -21.375951
[5,]   17.998493   6.002432
```

These, cv, are the canonical variates. We then form the canonical scores and plot them (see Figure 3.1, p. 24).

```
scores<-X %*% cv
plot(scores[,1],scores[,2],type="n",
     xlab="canonical variate 1",ylab="canonical variate 2")
```

```
text(scores[,1],scores[,2],crabs$sex, col=c(rep(4,100),rep(2,100)))
```

We can write a simple function my.lda.1 to implement the direct method of solving for the linear discriminants. It differs from what we have done here only in that it allows for a tolerance to determine whether Σ_W is singular. If any of the eigenvalues of Σ_W are less than the tolerance then a generalized inverse is used. In any production code it is a good idea to include such tests and if possible, work-arounds.

```
> leptograpsus.model.1<-my.lda.1(X, target, tol = 1e-09)
```

Another option is to use the code from the "Modern Applied Statistics with S" (MASS) library (Venables and Ripley, 2002). The lda function from this library has a large number of interesting options including some for a robust LDA. We compare the output from my.lda.1 with lda and find that they are the same except for sign. The LDA is only unique up to sign.

```
> library(mlp) #to load my.mlp.1
> leptograpsus.model.2<-lda(X,grouping=rep(1:4,each=50),method="moment")
> leptograpsus.model.1$scaling[,1]
[1]   31.217207    9.485303    9.822169 -65.950295   17.998493
> leptograpsus.model.2$scaling[,1]
        FL          RW          CL          CW          BD
-31.217207   -9.485303   -9.822169   65.950295  -17.998493
```

3.5 We consider a simulated problem designed to show off the features of quadratic discriminant analysis. This is the problem with data from two classes lying within two sphere, one inside of the other.

```
library(MASS)
set.seed(123)
data<-mvrnorm(200,c(0,0),matrix(c(1,0,0,1),2,2))

tt<-sqrt(data[,1]^2 + data[,2]^2)
sum(tt < 1.14)
dat1<-data[(tt < 1),]
dat2<-data[(tt > 1),]
temp<-rbind(dat1,dat2)
eqscplot(temp[,1],temp[,2],type="n")
text(temp[,1],temp[,2],c(rep(1,100),rep(2,100)))

symbols(0,0,circle=c(1.14),add=T)

data<-temp
xx<-as.matrix(cbind(data,data^2,data[,1]*data[,2]))
target<-make.target(c(100,100),low=0)

tt<-lda(xx,grouping=c(rep(1,100,),rep(2,100)))
ss1<-predict(tt)
```

```
a<-seq(-4,4,length=20)
aa<-expand.grid(a,a)
temp<-as.matrix(cbind(aa,aa^2,aa[,1]*aa[,2]))
dimnames(temp) <- NULL

ss<-predict(tt,newdata=temp)$posterior
ss<-matrix(ss,20,20)
contour(a,a,ss,add=T,levels=0.5)

beta<-solve(t(xx) %*% xx) %*% t(xx) %*% target
target.hat<-xx %*% beta
tt2<-lda(target.hat,grouping=c(rep(1,100,),rep(2,100)))
ss2<-predict(tt2)
a<-seq(-4,4,length=20)
aa<-expand.grid(a,a)
aa<-as.matrix(aa)
temp<-as.matrix(cbind(aa,aa^2,aa[,1]*aa[,2]))
dimnames(temp) <- NULL

aa.hat<-temp %*% beta
ss<-predict(tt2,newdata=aa.hat)$posterior
ss<-matrix(ss,20,20)
contour(a,a,ss,add=T,levels=0.5,col=2)


######### qda
tt2<-qda(data,grouping=c(rep(1,100,),rep(2,100)))
dimnames(aa) <- NULL
ss3<-predict(tt2,newdata=data)
ss<-predict(tt2,newdata=aa)$posterior[,1]
ss<-matrix(ss,20,20)

contour(a,a,ss,add=T,levels=0.5,col=3)


legend(1.4,-1.5,c("Bayes","qda","flexible  lda"),
       lty=1,col=c(1,3,2))
```

3.6   Using the data set of Problem 3.5 fit flexible discriminant models using the mda library (Hastie et al., 2006). The default it to do fit an LDA model to the fitted values from a linear model, however, method=mars and method=bruto will produce more interesting results.

3.7   Using the data set of Problem 3.5 fit an MLP model to both the observed and the extended data set.

3.8   The code to draw the ROC curve in Figure 3.6 is given in the scripts. Extend this to get ROC curves for the mlp and fda models.

3.9 In the space of the canonical variates, we can determine the Mahalanobis distance from each class mean for each point

$$D_q^2(x) = (x - \hat{\mu}_q)^T \hat{\Sigma}_g^{-1} (x - \hat{\mu}_q).$$

$D_q^2(x) \sim \chi(r)$ where r is the number of canonical variates. This tail area on the χ^2 distribution is referred to as the typicality probability (McLachlan, 1992) It may be the case that observations, that are put into a class with a high posterior probability, are found to have a low typicality.

Write a function to calculate the typicality and investigate the iris data set to see if all of the observations are typical of their class.

CHAPTER 4

ACTIVATION AND PENALTY FUNCTIONS

4.1 INTRODUCTION

We consider the conditions under which the outputs of an MLP may be interpreted as the posterior probabilities of class membership. We also consider the decomposition of the mean-squared error into a bias and a variance term. The relationship of the standard MLP model to log–linear models is discussed as well as some aspects of the choice of penalty and activation functions for classification tasks.

4.2 INTERPRETING OUTPUTS AS PROBABILITIES

We write X as a random variable, of which x is a realization, and we describe the classes as C_q for $q = 1, \ldots, Q$. We then write: $P(x)$ as the probability of observing a particular value of the variable X; $P(C_q, X)$ as the joint probability of observing X and the observation belonging to class C_q; and $P(C_q|X)$ as the conditional probability of the observation being in class C_q given that X was observed.

We consider the expected value of ρ_l over the support of the random variable X, dropping[1] the constant of $1/2$.

---

[1] The constant of $1/2$ is simply added by convention so that that constants cancel when the derivative is calculated.

A Statistical Approach to Neural Networks for Pattern Recognition by Robert A. Dunne   **35**
Copyright © 2007 John Wiley & Sons, Inc.

$$
\begin{aligned}
\mathbb{E}(\rho_l) &= \mathbb{E}_X\left[\mathbb{E}_{(t|X)}(\rho_l|X)\right] \\
&= \mathbb{E}\left[\sum_{q_1=1}^{Q}\left\{\mathbb{E}[z_{q_1}^{*\,2}|X] - 2\mathbb{E}[z_{q_1}^{*}t_{q_1}|X] + \mathbb{E}[t_{q_1}^{2}|X]\right\}\right] \\
&= \mathbb{E}\left[\sum_{q_1=1}^{Q}\left\{z_{q_1}^{*\,2} - 2z_{q_1}^{*}\mathbb{E}[t_{q_1}|X] + \mathbb{E}[t_{q_1}^{2}|X]\right\}\right].
\end{aligned}
$$

Adding and subtracting the term $\sum_{q_1=1}^{Q}\mathbb{E}^2[t_{q_1}|X]$ allows the above to be recast in the form

$$
\mathbb{E}(\rho_l) = \mathbb{E}\left[\sum_{q_1=1}^{Q}\left\{z_{q_1}^{*} - \mathbb{E}[t_{q_1}|X]\right\}^2\right] + \mathbb{E}\left[\sum_{q_1=1}^{Q}\mathbb{V}[t_{q_1}|X]\right]. \tag{4.1}
$$

Now as $\mathbb{V}[t_{q_1}|X]$, the conditional variance of the class labels t, does not depend on the inputs to the network, we only need to minimize the first term. It can be seen that this term is minimized when the network gives a least squares estimate of the expected value of the targets, conditional on the observations X.

If we consider target values with a "one of Q" encoding (Section 2.1, p. 9) then, as

$$
P(t_{q_1}|X) = \sum_{q_2=1}^{Q} P(t_{q_1}|C_{q_2})\,P(C_{q_2}|X)
$$

and $P(t_{q_1}|C_{q_2}) = \delta(t_{q_1} - \delta_{q_1 q_2})$, we can argue that $\mathbb{E}(\rho)$ is minimized when

$$
\begin{aligned}
z_{q_1}^{*}(X) &= \mathbb{E}[t_{q_1}|X] \\
&= \int t_{q_1} P(t_{q_1}|X)dt_{q_1} \\
&= \int t_{q_1} \sum_{q_2=1}^{Q} \delta(t_{q_1} - \delta_{q_1 q_2})P(C_{q_2}|X)dt_{q_1} \\
&= P(C_{q_1}|X). \tag{4.2}
\end{aligned}
$$

Hence ρ is minimized when the network outputs are equal to the posterior probability of class membership[2].

That the MLP directly estimates the *a posteriori* probability $P(C_q|X)$ when the targets are "one of Q" and an appropriate objective function is used, has appeared a number of times in the literature. Richard and Lippmann (1991) point out that it is true for the sum of squares and cross-entropy penalty functions. Hampshire and Perlmutter (1990) consider the general conditions on ρ under which the network outputs may be interpreted as probabilities. They show that if a function e satisfies

$$
e(z^{*}) = \int (z^{*})^{r}(1 - z^{*})^{r-1}dz^{*},
$$

---

[2]Note that $\delta_{q_1 q_2}$ is the Kronecker delta, while $\delta(x)$ is the Dirac delta.

assuming the "one of Q" encoding, then e will give the functional form of a ρ that will produce consistent estimators of the posterior probability of class membership. For $r = 1$, the sum of squares penalty function is recovered, while for $r = 0$ cross-entropy is recovered.

In the case of targets other than $\{0, 1\}$, for example $\{0.1, 0.9\}$ targets, the outputs will not be the posterior probabilities (Pao, 1989). Consider a two class problem, C_1 and C_2, with target values a and b respectively. Then

$$z^* = aP(C_1|X) + bP(C_2|X)$$

(Gish, 1990) and similarly for multi–class problems. Hence the outputs are not directly interpretable as posterior probabilities; however, a set of linear equations could be solved to recover the probabilities (provided there are as many outputs as classes).

4.3   THE "UNIVERSAL APPROXIMATOR" AND CONSISTENCY

It is known that MLPs with sigmoid or linear output units, and any of a range of "squashing" activation functions at the hidden layer, are universal approximators on closed, bounded sets. That is, given a continuous $f : \mathbb{R}^p \to \mathbb{R}^q$ there is a z^*, output by an MLP, such that for closed, bounded $K \subset R^p$ and $\epsilon > 0$

$$\|f(x) - z^*(x)\| < \epsilon \quad \forall x \in K.$$

This result appears in Hornik et al. (1989) and Funahashi (1989); see also Baldi (1991). In order to achieve this result the number of hidden–layer units must be allowed to grow without bound. There are also results for discontinuous functions, see Ripley (1996, §5.7) and Scarselli and Tsoi (1998) for surveys of the area and further references.

Uniform L^2 convergence is a strong result and implies $L^2(\mu)$ convergence where μ is some probability measure on $\mathbb{R}^p$ (Halmos, 1974). Thus the result can be specialized to convergence in probability and consistency. Also there are results in the literature showing, with varying conditions, that $z^*(D, x)$ is a pointwise mean square consistent estimator of $\mathbb{E}_D[T|x]$ so that

$$\lim_{N \to \infty} \mathbb{E}_D[(z^*(D, x) - \mathbb{E}_D[T|x])^2] = 0$$

or

$$z_q^* \to P(C_q|X) \text{ as } N \to \infty$$

(White, 1989; Mielniczuk and Tyrcha, 1993; Geman et al., 1992). Hampshire and Perlmutter (1990) specialize this result to show that an MLP classifier approaches the Bayesian decision boundary with a wide variety of penalty functions, including those that do not allow the MLP outputs to be interpreted as probabilities.

A similar result for the softmax activation function (Equation (2.3), p. 11) does not appear in the literature, but can be seen to follow if we consider a softmax output for a particular class to be equivalent to a logistic activation function for that class versus all of the other classes.

There are two different aspects here:

1. Equations (4.1) and (4.2) show that ρ is minimized when $z_q^* = P(C_q|X)$;

2. The results cited above show that this minimum is reached under certain conditions. That is, $z_q^* \rightarrow P(C_q|X)$ as $N \rightarrow \infty$ provided the MLP is allowed to have an arbitrary number of hidden layer units.

However, allowing the number of hidden layer units to be arbitrarily large for a fixed set of training data is not a good strategy. As Richard and Lippmann (1991) point out, the estimates z_q^* of $P(C_q|X)$ will only be good if:

1. sufficient training data are available;

2. the network has sufficient complexity to implement the correct classification;

3. the classes are sampled with the correct *a priori* distribution.

Note that a combination of logistic outputs with a sum of squares penalty function gives unbiased estimators of the posterior probability of class membership, $P(C|X)$, given the conditions above. Hence we have the asymptotic property that the outputs from a multi–class MLP with a sum of squares and logistic functions will also sum to one. However, these are asymptotic results and may not help greatly in dealing with small samples. Note that nothing in this discussion depends on the fact that we are fitting an MLP model. All that is assumed is that the method is flexible enough to model $P(C \mid x)$.

4.4   VARIANCE AND BIAS

Suppose we have a training set, D_N, consisting of N sample points. We consider the left-hand side of equation (4.1), but now we take the expectation over D_N. We consider the training set to have been drawn from a population of training sets of size N, and to emphasize this we write $z^*(D_N, x)$; z^* estimated from D_N and evaluated at the point x. Suppressing the summation over q we write (using the same argument as in the derivation of (4.1)),

$$\mathbb{E}_{D_N}[(z^*(D_N, x) - \mathbb{E}_{D_N}[T|x])^2] = \mathbb{E}_{D_N}\left[\left(z^*(D_N, x) - \mathbb{E}_{D_N}[z^*(D_N, x)]\right)^2\right]$$
$$+ \left(\mathbb{E}_{D_N}[z^*(D_N, x)] - \mathbb{E}_{D_N}[T|x]\right)^2$$
$$= \text{variance} + \text{bias}^2.$$

There are a number of forms of consistency, depending on the type of convergence that is considered; however, any reasonable definition will require that both bias and variance go to zero as the size of D_N is increased. The derivation of consistency results for the MLP can be described heuristically (Geman et al., 1992) as the process of decreasing the bias, by letting the size of the network grow as $N \rightarrow \infty$, but not too fast so that the variance does not grow too fast.

However for finite D there is a trade–off between bias and variance. As one fits models with more parameters the bias is decreased; however, the variance of the estimator z^* is increased (Geman et al., 1992). Hence there is an optimal point beyond which adding more parameters is counterproductive. Unfortunately in practice estimating this point is difficult and computationally expensive. This is essentially the problem of overfitting and in many instances the best practical advice that can be offered is simply to limit the number of parameters in an arbitrary fashion. See Section 5.3.5 (p. 61) for a discussion of the Network Information Criteria and other methods for model selection (see also Hastie et al., 2001, Section 7 for a discussion of this).

4.5 BINARY VARIABLES AND LOGISTIC REGRESSION

Minimizing the sum of squares penalty function coincides with finding the maximum likelihood estimator for a random variable with a Gaussian distribution.

However, in a classification task, as the target T takes values in the set $\{0,1\}$, the Gaussian distribution is not a plausible model. An alternative is to assume that T is a Bernoulli random variable and let $\pi = P(T = 1)$ and $1 - \pi = P(T = 0)$. We can then write

$$p(T|\pi) = (\pi)^T (1 - \pi)^{1-T}, \tag{4.3}$$

which is the likelihood function for a Bernoulli random variable.

The Bernoulli distribution can be written as a member of the exponential family of distributions[3]

$$p(T|\theta, \psi) = \exp\left\{\frac{\theta^T T - b(\theta)}{a(\psi)} + c(T, \psi)\right\}, \tag{4.4}$$

where θ is a location parameter and ψ is a scale parameter. The fact that T has an exponential family distribution means that a well developed body of theory, that of *generalized linear models* (McCullagh and Nelder, 1989), can be applied. Write $\mu = \mathbb{E}(T)$, then in a linear model $\hat{\mu} = X\omega$, and the ω are estimated by least squares. In a generalized linear model we write $\eta = X\omega$, where η is called the linear predictor, and then model $\eta = g(\mu)$, where g is the link function. The link function for which $\theta = \eta$ is referred to as the canonical link. If p is a Gaussian distribution then $\theta = \mu$ and $\psi = \sigma^2$ and if g is the identity function we have $X\omega = \eta = \mu$, the usual linear model.

To write (4.3) in the form of equation (4.4), we reparameterize it as

$$p(T|\pi) = \exp\left\{T\log\left(\frac{\pi}{1-\pi}\right) + \log(1-\pi)\right\}, \tag{4.5}$$

where $\mathbb{E}(T) = \pi$ and $\theta = \log(\pi/(1-\pi))$. Thus the canonical link for the Bernoulli distribution is $\log(\pi/(1-\pi))$, and $\pi = 1/(1+\exp(-\theta))$, and so the logistic function is recovered as the inverse canonical link. Extending the model (4.3), say we have a sample of size m and let $Y = \sum_{i=1}^{m} T_i$. Y is then a binomial random variable with distribution

$$P(Y = y) = \binom{m}{y} \pi^y (1 - \pi)^{m-y}$$

---

[3]The exponential family of distributions includes the binomial, Poisson, negative binomial, gamma, Gaussian, uniform, geometric, exponential etc. See Silvey (1975) and Rao (1973) for discussions of the properties.

for $y = 0, 1, \dots, m$. Using (4.5), we can write the binomial distribution as a member of the exponential family.

Say we model π via $g(\pi) = \omega_0 + \omega_1 x$, so that we have a generalized linear model with link function g. To ensure that $\pi \in [0, 1]$, we model the inverse of g as a cumulative probability distribution

$$\pi = g^{-1}(\omega_0 + \omega_1 x) = \int_{-\infty}^{x} f(s)ds.$$

$f(s)$ is termed the "tolerance" distribution and, for example, the probit model chooses f as the Gaussian distribution. The logistic chooses

$$f(s) = \frac{\omega_1 \exp(\omega_0 + \omega_1 s)}{\{1 + \exp(\omega_0 + \omega_1 s)\}^2}$$

so that

$$\pi = \int_{-\infty}^{x} f(s)ds$$
$$= \frac{\exp(\omega_0 + \omega_1 x)}{1 + \exp(\omega_0 + \omega_1 x)}.$$

Say, furthermore, that we have observed a number of m_n and y_n at discrete values of the explanatory variables X so that $g(\pi_n) = \omega^T x_n$. Note that the columns of X may be the values of a continuous variable, or levels of a factor variable. Fitting the binomial distribution then leads to maximizing the log likelihood

$$\ell(\pi, y) = \sum_{n=1}^{N} \left\{ y_n \log \pi_n + (m_n - y_n) \log(1 - \pi_n) + \log \binom{m_n}{y_n} \right\},$$

where

$$\hat{\pi}_n = \frac{\exp(\omega^T \mathbf{x}_n)}{1 + \exp(\omega^T \mathbf{x}_n)},$$

with respect to the ωs. The logistic model is fitted by forming $\frac{\partial \ell}{\partial \omega}$ and $\frac{\partial^2 \ell}{\partial \omega^2}$ and using the "iteratively re-weighted least squares" algorithm (Appendix A.4, p. 249) to estimate the parameters of the model.

4.6   MLP MODELS AND CROSS–ENTROPY

In the case of the MLP model, we take each x_i to be unique, so that $N_i = 1$ and each $y_i = 0$ or 1. To estimate the parameters ω for the MLP model, we form the likelihood

$$L = \prod_{n=1}^{N} (z_n^*)^{t_n} (1 - z_n^*)^{1 - t_n},$$

and then the negative log likelihood

$$\mathcal{L} = -\log(L) = -\sum_{n=1}^{N} \{t_n \log(z_n^*) + (1 - t_n) \log(1 - z_n^*)\}. \qquad (4.6)$$

Differentiating with respect to the parameter ω_p, we get

$$
\begin{aligned}
\frac{\partial \mathcal{L}}{\partial \omega_p} &= \sum_{n=1}^{N} \frac{\partial \mathcal{L}}{\partial z_n^*} \frac{\partial z_n^*}{\partial \omega_p} \\
&= \sum_{n=1}^{N} (t_n - z_n^*) x_p
\end{aligned}
$$

and then a gradient–based scheme can be used to minimize the negative log likelihood.

Some slight extensions are necessary when dealing with multiple, mutually–exclusive classes. For an MLP with Q outputs and a "one of Q" target encoding, we write

$$
p(t|x) = \prod_{q=1}^{Q} (z_q^*)^{t_q}.
$$

Taking the negative log likelihood for a single observation gives a penalty function of the form

$$
\rho' = -\sum_{q=1}^{Q} t_q \log(z_q^*),
$$

and subtracting the minimum of ρ' gives

$$
\begin{aligned}
\rho &= \rho' - \rho_{\min} \\
&= -\sum_{q=1}^{Q} t_q \log(z_q^*/t_q) = \rho_c \text{ say,}
\end{aligned}
\tag{4.7}
$$

which is known as the cross–entropy penalty function. This penalty function is non–negative and zero at the minimum, which occurs when $z_q^* = t_q$ for all q outputs, provided the constraint

$$
\sum_q z_q^* = 1
\tag{4.8}
$$

is satisfied. See Section 4.8.1 (p. 45) for a discussion of the necessity of this condition.

It can be seen that there are a number of connections between the MLP model and other statistical models (Sarle, 1994). The standard MLP with an identity activation function, no hidden–layer and a sum of squares penalty function will give the usual linear regression model. The MLP model of size $P.1$ with ρ given by (4.6), f_Q chosen as a logistic activation function and $t \in \{0, 1\}$ is identical to the logistic regression model. We further note that, with some parameter redundancy, this is the same model as $P.2$ with ρ given by (4.7), f_Q chosen as the softmax function and t taking the values $(0, 1)$ and $(1, 0)$. This model can be extended to one of size $P.Q$ which will fit a multinomial model (as discussed in the next section). However, MLP models of size $P.H.Q$ are not represented in the statistics literature, although they can be seen (Ripley, 1996, §5.2) to have some similarity to projection-pursuit models.

4.6.1 Contingency tables

Contingency tables and log-linear models have an extensive literature, see McCullagh and Nelder (1989) and Dobson (1990) for treatments of the area. Restricting attention to a two–way table, suppose that there are two factor variables A, with levels $j = 1, \ldots, J$, and B, with levels $k = 1, \ldots, K$, giving a two–dimensional grid of size $J \times K$. For each cell in the grid there is an associated number, y_{jk}, of observations.

The aim here is to construct the simplest model that is consistent with the data. The data can be modeled as a product of JK Poisson distributions with means $\lambda = \exp(X\omega)$, where X is a matrix with factor entries to indicate the cells of the contingency table, so that

$$f(y|\lambda) = \prod_{j,k} \frac{\lambda_{jk}^{y_{jk}} \exp(-\lambda_{jk})}{y_{jk}!}.$$

Alternatively, if $n = \sum_{j,k} y_{jk}$ is taken as fixed in advance, a multinomial model can be fitted,

$$
\begin{aligned}
f(y|n, \lambda) &= \prod_{j,k} \frac{\lambda_{jk}^{y_{jk}} \exp(-\lambda_{jk})}{y_{jk}!} \frac{n!}{\lambda_{..}^{n} \exp(-\lambda_{..})} \\
&= n! \prod_{j,k} \frac{\theta_{jk}^{y_{jk}}}{y_{jk}!},
\end{aligned}
$$

where $\alpha_{.}$ means that α_i has been summed over the index i, and $\theta_{jk} = \dfrac{\lambda_{jk}}{\lambda_{..}}$ is the probability for a given cell. In addition there are product multinomial models that involve fixed row or column totals. Although we may be interested in one of the multinomial models, it is easier to fit the Poisson model. Such a model is called a "surrogate Poisson" model and, with correctly selected model terms, will give the same estimates (Birch, 1963) as the multinomial model.

Where one of the factors has two levels, a binomial model may be fitted, giving the same parameter estimates (see Venables and Ripley (1994, §7.3) for an example). This can be extended to a response factor with more than two levels by using an MLP with appropriate activation and penalty functions. In such a case an MLP with no hidden layers is fitting a multinomial model without a surrogate Poisson model[4]. Say that for the response factor we have three levels and, for a particular cell, we have the following counts, (y_1, y_2, y_3), for the three levels. Then the target vector is $(y_1/y_., y_2/y_., y_3/y_.)$ and the penalty function is weighted by $y_.$. In other words, the targets are observed probabilities and the MLP models these probabilities. A hierarchy of nested models may be fitted, from the saturated model (with a separate term for each cell) to the intercept model and the final model may be selected via the AIC criterion. We provide a brief discussion of the AIC criterion and further references in Section 5.3.5, (p. 61).

---

[4]See Ripley (1994b) for an S package ("multinom") that allows such a hierarchy of contingency table models to be fitted via single-layer MLP models.

4.7 A DERIVATION OF THE SOFTMAX ACTIVATION FUNCTION

The softmax activation function

$$z_q^* = \frac{\exp(z_q)}{\sum_{q_1} \exp(z_{q_1})}$$

ensures the condition of equation (4.8) is met and allows the use of the cross–entropy penalty function. However, by modeling $P(x|C)$ we can give a better justification for the use of the softmax activation function. For a classification scheme using the sampling paradigm, for a two-class problem, $P(C_1|x)$ may be modeled as

$$P(C_1|x) = \frac{P(x|C_1)P(C_1)}{P(x)}$$

$$= \frac{P(x|C_1)P(C_1)}{P(x|C_1)P(C_1) + P(x|C_2)P(C_2)},$$

which can be written as

$$= \frac{1}{1 + \exp\left\{ -\log\left[\frac{P(x|C_1)}{P(x|C_2)}\right] - \log\left[\frac{P(C_1)}{P(C_2)}\right]\right\}}. \qquad (4.9)$$

Now, if we are making some distributional assumptions about $P(x|C_q)$, it is a standard procedure to base a test of C_1 versus C_2 on the likelihood ratio,

$$\text{LR} = \frac{P(C_1|x)}{P(C_2|x)}.$$

For computational reasons, we take minus the log of the likelihood and minimize this with respect to the parameters of the distribution $P(x|C_q)$.

$$\mathcal{L} = -\log(\text{LR})$$

$$= -\log\left[\frac{P(x|C_1)}{P(x|C_2)}\right] - \log\left[\frac{P(C_1)}{P(C_2)}\right].$$

If, for example, we assume that the classes have Gaussian distributions with a common covariance matrix Σ, and means μ_1 and μ_2

$$\mathcal{L} = \frac{1}{2}\left\{-(x-\mu_1)^T \Sigma^{-1}(x-\mu_1) + (x-\mu_2)^T \Sigma^{-1}(x-\mu_2)\right\} - \log\left[\frac{P(C_1)}{P(C_2)}\right]$$

$$= \frac{1}{2}\left\{-x^T\Sigma^{-1}x + 2\mu_1^T\Sigma^{-1}x + \mu_1^T\Sigma^{-1}\mu_1 + x^T\Sigma^{-1}x - 2\mu_2^T\Sigma^{-1}x - \mu_2^T\Sigma^{-1}\mu_2\right\}$$

$$\qquad - \log\left[\frac{P(C_1)}{P(C_2)}\right]$$

$$= \alpha + \beta^T x$$

where $\beta^T = (\mu_1 - \mu_2)^T\Sigma^{-1}$ and

$$\alpha = \frac{1}{2}\left[(\mu_1 + \mu_2)^T\Sigma^{-1}(\mu_1 + \mu_2)\right] - \log\left\{\frac{P(C_1)}{P(C_2)}\right\}$$

subsumes the constant terms. Hence in this case the posterior probabilities can be written as a logistic function of a linear combination of the variables. In this treatment the logistic function only arises as a convenient mathematical step. Jordan (1995) comments that (4.9) will only be useful if the log likelihood has some convenient and tractable form, as it does in the example above.

However, if we start with multiple classes and assume that $P(X|C_q)$ is a distribution from the exponential family of distributions parameterized by (θ_q, ψ), we can derive the softmax activation function directly. Note that the distributions are assumed to have a common scale ψ.

$$
\begin{aligned}
P(C_{q_1}|X) &= \frac{P(X|C_{q_1})P(C_{q_1})}{\displaystyle\sum_{q_2=1}^{Q} P(X|C_{q_2})P(C_{q_2})} \\[2mm]
&= \frac{P(X|\theta_{q_1},\psi)P(C_{q_1})}{\displaystyle\sum_{q_2=1}^{Q} P(X|\theta_{q_2},\psi)P(C_{q_2})} \\[2mm]
&= \frac{\exp\left\{\dfrac{\theta_{q_1}^{\tau}X - b(\theta_{q_1})}{a(\psi)} + c(X,\psi)\right\}P(C_{q_1})}{\displaystyle\sum_{q_2=1}^{Q}\exp\left\{\dfrac{\theta_{q_2}^{\tau}X - b(\theta_{q_2})}{a(\psi)} + c(X,\psi)\right\}P(C_{q_2})} \\[2mm]
&= \frac{\exp\left[\dfrac{\theta_{q_1}^{\tau}}{a(\psi)}X - \dfrac{b(\theta_{q_1})}{a(\psi)} + \log\{P(C_{q_1})\}\right]}{\displaystyle\sum_{q_2=1}^{Q}\exp\left[\dfrac{\theta_{q_2}^{\tau}}{a(\psi)}X - \dfrac{b(\theta_{q_2})}{a(\psi)} + \log\{P(C_{q_2})\}\right]}.
\end{aligned}
\tag{4.10}
$$

Note that $\{a(\psi)\}^{-1}\theta_{q_1}^{\tau}X - \{a(\psi)\}^{-1}b(\theta_{q_1}) + \log\{P(C_{q_1})\}$ is a linear combination of the variables with an offset or bias term and that (4.10) is the softmax activation function. This shows that modeling the posterior as a softmax function is invariant to a family of classification problems where the distributions are drawn from the same exponential family with equal scale parameters. The logistic activation function is then recovered as a special case of softmax.

4.8   THE "NATURAL" PAIRING AND Δ_Q

We consider the case where z_q^* is given by the softmax activation function and ρ_c is the cross–entropy activation function. Then

$$
\begin{aligned}
\frac{\partial \rho_c}{\partial z_{q_1}} &= \sum_{q_2=1}^{Q} \frac{\partial \rho_c}{\partial z_{q_2}^*} \frac{\partial z_{q_2}^*}{\partial z_{q_1}} \\
&= \sum_{q_2=1}^{Q} \frac{-t_{q_2}}{z_{q_2}^*} (z_{q_2}^* \delta_{q_1,q_2} - z_{q_1}^* z_{q_2}^*) \\
&= \sum_{q_2=1}^{Q} -t_{q_2} (\delta_{q_1,q_2} - z_{q_1}^*) \\
&= \sum_{q_2=1}^{Q} (t_{q_2} z_{q_1}^* - \delta_{q_1,q_2} t_{q_2}) \\
&= \left(\sum_{q_2=1}^{Q} t_{q_2} \right) z_{q_1}^* - t_{q_1} \\
&= \Delta_{q_1} \text{ say.}
\end{aligned}
\tag{4.11}
$$

Hence, for $v_{hq} \in \Upsilon$, we can write

$$
\begin{aligned}
\frac{\partial \rho_c}{\partial v_{hq}} &= \sum_{q_1=1}^{Q} \frac{\partial \rho_c}{\partial z_{q_1}^*} \frac{\partial z_{q_1}^*}{\partial z_q} \frac{\partial z_q}{\partial v_{hq}} \\
&= \Delta_q \frac{\partial z_q}{\partial v_{hq}}
\end{aligned}
$$

and we can go on to calculate the derivatives of the other layers of the MLP.

Bishop (1995a, §6.8) gives Δ_q as $(z_q^* - t_q)$. Since $\sum_{q_1=1}^{Q} t_{q_1}$ will, in general, equal one, (4.11) and Bishop's Δ_q will coincide. Bishop suggests that as:

- identity output units and a least squares penalty function;

- a two–class cross–entropy penalty function and a logistic activation function; and

- a multi–class cross–entropy penalty function and a softmax activation function

all give the same $\Delta_q = z_q^* - t_q$, there is a natural pairing of activation functions and penalty functions. When the natural pairing is used, it will always be the case that $\Delta_q = z_q^* - t_q$. With the framework of generalized linear models (Section 4.5, p. 39), these choices of activation functions can be seen to be canonical link functions corresponding to the distribution indicated by the choice of penalty function, respectively the Gaussian, Bernoulli and multinomial distributions.

4.8.1   Constraints needed for cross–entropy

The cross–entropy and sum of squares penalty functions are plotted for t and z^* in the region $[0,1]^2$ in Figures 4.1 and 4.2. For the sum of squares function, ρ has a minimum value of 0 whenever $z^* = t$. However, for the cross–entropy penalty function, there is a minimum of $-e^{-1}$ at the point $(z^*, t) = (1, e^{-1})$ and, for a given non–zero target t_q, the function is minimized when $z_q^* = 1$. For a zero target, the cross–entropy penalty function is minimized for any value of z^*. Hence the MLP may simply return 1 for all classes. It is thus only the additional constraint that $\sum_q z_q^* = 1$ which makes the cross–entropy penalty function usable.

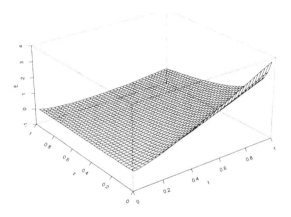

Figure 4.1   The cross–entropy penalty function $\rho = t \log(t/z^*)$. The function has a minimum of $-e^{-1}$ at the point $(z^*, t) = (1, e^{-1})$ and, for a given non–zero target t, the function is minimized when $z^* = 1$. The function is undefined at $t = 0$ and ∞ at $z^* = 0$.

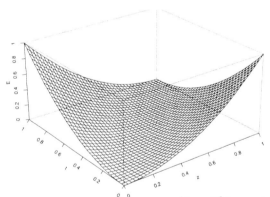

Figure 4.2   The least squares penalty function $\rho = \frac{1}{2}(t - z^*)^2$ has a minimum value of 0 when $z^* = t$.

ρ_c is a finite sample version of the Kullback–Liebler divergence (or relative entropy)

$$d(p_1, p_2) = \int p_1 \log \left(\frac{p_1}{p_2} \right) dx$$

between two probability distributions p_1 and p_2. It is not symmetric and is thus not a distance metric. When ρ_c is derived in this light, it is clear that p_1 and p_2 should have the properties of distributions, in particular that they should integrate to 1. However, a reading of some of the standard references on this (Bridle, 1990; Bishop, 1995a) throws little light on the necessity[5] of the condition that $\sum_q z_q^* = 1$.

The sum of squares penalty function can also be paired with the softmax activation function. Table 4.1 shows the Δ_q term for each of the possibilities. Note that the combination of sum of squares and softmax gives a more complex Δ_q.

Table 4.1 Δ_q for various pairings of penalty and activation functions.

	least squares	cross-entropy
logistic	$(z_q^* - t_q)(z_q^* - (z_q^*)^2)$	$z_q^* - t_q$
softmax	$(z_q^*)^2 - t_q z_q^* + z_q^*(\sum_{k_1} t_{k_1} z_{k_1}^* - (z_{k_1}^*)^2)$	$z_q^* - t_q$
identity	$z_q^* - t_q$	$\dfrac{-t_q}{z_q^*}$

4.9 A COMPARISON OF LEAST SQUARES AND CROSS–ENTROPY

Leaving aside the identity link as being inappropriate for a classification task, this still leaves us with three possible estimators to consider:

1. least–squares and the logistic activation function;

2. least–squares and the softmax activation function;

3. cross–entropy and the softmax activation function.

The justifications for using these particular penalty functions are quite disparate. Both have been shown to result in z_q^* approximating $P(C_q|x)$ when used with the "one of Q" target encoding. Least squares is justified on the basis of the Gauss–Markov theorem or intuitively, while cross–entropy is derived as a maximum likelihood (ML) estimator by modeling t_{nq} as a Bernoulli random variable. This gives us the result that, with some regularity conditions, if there exists an unbiased estimator which attains the Cramér–Rao minimum variance bound, then the ML estimator coincides with it. This is not helpful with an MLP model, as the point estimate of the posterior probability $P(C|x)$ can only be shown to be unbiased when there are an arbitrary number of hidden–layer units (Hornik et al., 1989). The practice with very flexible models like the MLP is to introduce some bias in order to reduce the variance of the estimates (Geman et al., 1992). In addition, the properties of Fisher efficiency that many ML estimators enjoy does not follow automatically, but has to be shown in each particular case (Silvey, 1975).

Why then would we use one estimator rather than another? In particular, we would like to know how fast z_q^* converges to $P(C_q|x)$ (is it useful for reasonable–

[5] The author mistakenly implemented an MLP with ρ_c and a logistic activation function before realizing the problems with it.

sized samples?) and how variable the two estimates are. There appears to be little guidance in the literature on these questions.

4.10 CONCLUSION

Within the framework of function approximation theory (Hornik et al., 1989), all that is required to show that the MLP is a universal approximator is that the activation functions be smooth, bounded and monotonic. In addition, it can be seen that a number of different penalty terms lead to $z_q^* \to P(C_q|X)$, and that a wider class of penalty functions lead to the Bayes decision rule.

While the experiment described here is not a substitute for a theoretical understanding of the properties of the three estimators, within its limitations it does suggest that a probabilistic approach to the MLP model may yield more accurate results. That is, modeling $P(x|C)$ as an exponential family distribution, and the target values as Bernoulli random variables, thus recovering the softmax activation function and the cross–entropy penalty function, may lead to more accurate results for a finite sample size than other activation and penalty function pairs chosen, on non–probabilistic grounds, to meet the conditions for the MLP to be a universal approximator.

That this should be so in the case of the logistic versus the softmax estimates of $P(C_q|x)$ is not surprising, as the logistic outputs fail to sum to one in some regions of the feature space. However, it is more surprising that the classification accuracy also appears to be improved with the cross-entropy penalty function and the softmax function.

4.11 COMPLEMENTS AND EXERCISES

4.1   In the absence of any theoretical guidance, we consider a simple simulation. We take 100 observations from each of four Gaussian classes with means at the four symmetric points $\{\pm 1, \pm 1\}$ and identity covariance matrix, and conduct 100 trials, generating independent samples for the training and testing cycles.

We consider two aspects here. One is the accuracy of the estimates of $P(C|x)$ and the other is the class–conditional error rates. We measure the accuracy of $P(C|x)$ by the integrated difference between the estimated posterior probabilities and the exact probabilities

$$\sqrt{\sum_{q=1}^{4} \int \{\hat{P}(C_q|x) - P(C_q|x)\}^2 F(x)dx}. \tag{4.12}$$

We can calculate $P(C_q|x)$ as the distributions are known and we can estimate (4.12) by

$$\sqrt{\sum_{q=1}^{4} \frac{1}{N} \sum_{n=1}^{N} \{\hat{P}(C_q|x_n) - P(C_q|x_n)\}^2}. \tag{4.13}$$

The number of independent trials should ensure that (4.13) is an adequate estimator of (4.12) and allow us to rank the classifiers.

We can see from Table 4.2 that the estimators are ranked, from worst to best, in the order:

1. logistic activation function and least squares;

2. softmax activation function and least squares;

3. softmax activation function and cross-entropy.

It would appear that the improvement is due to both the activation function and the penalty function. The differences between the two softmax activation function models and the logistic model are significant at the 5 % level while the differences between the two softmax activation function models are not significant. We next calculated the class–conditional error rates using the known distributions (Table 4.3), and the empirical results (Tables 4.4, 4.5 and 4.6). It can be seen (by inspection or by a summary of the tables such as the trace) that the estimators are again ranked in the same order.

Table 4.2   A comparison of the estimates of the quantity (4.12) for the three estimators.

estimator	mean	variance
logistic activation function and least squares	4.621323	0.3146437
softmax activation function and least squares	4.003461	0.2315692
softmax activation function and cross-entropy	3.936804	0.1958714.

Table 4.3   The Bayesian class–conditional classification rates. The true class is shown down the table and the ascribed class across the table.

	1	2	3	4
1	0.7079	0.1335	0.1335	0.0252
2	0.1335	0.7079	0.0252	0.1335
3	0.1335	0.0252	0.7079	0.1335
4	0.0252	0.1335	0.1335	0.7079

In this example it would appear that even for reasonably large sample sizes (400 observations in 4 classes) the combination of the softmax activation function and the cross–entropy penalty function gives the more accurate result. However, the scope of the simulation is quite limited as the data is quite unstructured and well behaved (there are no outliers). For a further example (with both structured data and outliers) see experiment 4 in Chapter 9 (p. 143).

Redo the experiment but with a more challenging data set. One option is to add a percentage of observations coming from another distribution with fatter tails. This

Table 4.4   The class–conditional classification rates for the least squares error function and the logistic activation function. The true class is shown down the table and the ascribed class across the table.

	1	2	3	4
1	0.7001	0.1387	0.1336	0.0276
2	0.1450	0.6824	0.0243	0.1483
3	0.1350	0.0289	0.6999	0.1362
4	0.0344	0.1383	0.1398	0.6875

Table 4.5   The class–conditional classification rates for the cross–entropy penalty function and the softmax activation function. The true class is shown down the table and the ascribed class across the table.

	1	2	3	4
1	0.7046	0.1376	0.1318	0.0260
2	0.1333	0.6981	0.0246	0.1440
3	0.1302	0.0265	0.7055	0.1378
4	0.0297	0.1343	0.1335	0.7025

Table 4.6   The class–conditional classification rates for the least squares penalty function and the softmax activation function. The true class is shown down the table and the ascribed class across the table.

	1	2	3	4
1	0.7068	0.1397	0.1274	0.0261
2	0.1373	0.6958	0.0241	0.1428
3	0.1310	0.0261	0.7039	0.1390
4	0.0317	0.1349	0.1324	0.7010

device, modelling data as coming from $(1 - \epsilon)F_1 + \epsilon F_2$, where the F_2 distribution is a contaminating distribution, is common in the modeling of data with outliers. A good choice for F_2 is the Cauchy distribution.

4.2   The MASS library provides a birth weight data set of 189 observations and 10 variables. The data were collected at Baystate Medical Center, Springfield, MA, during 1986.

A number of the variables are categorical or factor variables. These have to be encoded numerically in order to fit the model. Examine the encoded model matrix to see how this is done.

Using the multinom function fit a log-linear model to the birth weight data to test for association between birth weight and the the 9 explanatory variables.

4.3   the nnet allows a single layer MLP to be fitted by specifying size=0 and skip=TRUE. Fit a single layer MLP to the design matrix from Exercise 4.2.

4.4   Using the binomial family fit a GLM model to the birth weight data set.

4.5   Dobson (1990) gives an ulcer data set (given in the script files). Fit a Poisson GLM and a `multinom` model to this data set.

CHAPTER 5

MODEL FITTING AND EVALUATION

5.1 INTRODUCTION

We consider the general question of the evaluation of a classifier and the questions of model selection and penalized fitting procedures for the MLP model.

MLP models are rarely used in circumstances where a low value of ρ at the observed data, D, is the desired goal. A central question in fitting and evaluating an MLP model is that of "generalization." The term comes from the original biological motivation of neural networks and to generalize well means to have the capacity to apply learned knowledge to new situations. Quantitatively it can be defined as the expected performance of the MLP, measured by ρ or by a misclassification rate or matrix. In practice the generalization ability of the MLP is often measured by the performance on a set of previously unseen data.

The MLP is a powerful (Section 4.3, p. 37) distribution–free regression method. However, this very power means that it is possible to reduce ρ to 0 on the training data. Now, if we consider the data to have been generated by a process with a systematic and a random component then, in order to reduce ρ to 0 the MLP must be modeling noise, and in this case it is likely that the MLP will not generalize well. The term "overfitting" is used to describe learning the random component in the training set. Weigend (1994) suggests that, as an operational definition, overfitting can be said to occur when the error on an independent test set, having passed through a minimum, starts to increase (see item 8, p. 65).

A Statistical Approach to Neural Networks for Pattern Recognition by Robert A. Dunne   **53**
Copyright © 2007 John Wiley & Sons, Inc.

Fitting a model to data involves negotiating a trade–off between bias and variance, as a model with either an inflated bias or variance will lead to an inflated mean squared error (see Section 4.4, p. 38; Geman et al. 1992; and Bishop 1995a §9). Such a model may not perform well on new data. For models with few parameters and strong distributional assumptions, such as linear models, model selection can be viewed as bias reduction. Such models are referred to as "confirmatory" and rely on the value of ρ on the training data to validate the model. In some circumstances, such as with (near) collinearities in the variables, penalized fitting procedures are necessary.

For flexible models, like the MLP, that can model a large class of densities given sufficient hidden–layer units, model selection and penalization are not clearly separable. It is rare that the "correct" MLP model can be selected in any meaningful sense. A more likely situation is that, whatever the correct model which defines the process generating the data is, the MLP is flexible enough to model this density. Because of this it is often the case that the MLP model will fit the training data too accurately, resulting in an overfitted model with a large variance. Given this, it is not surprising that linear models only rarely require penalized training, whereas MLP models almost invariably do.

5.2   ERROR RATE ESTIMATION

Let C represent the true class for X drawn from the population. While in general x is classified as belonging to $\arg\max_c P(c|x)$, it may be desirable to put x into a "doubt" class if $\max_c P(c|x)$ is below some specified value. The performance of a model may be assessed by a measure of the misclassification rate. Let

$$1 - P(\text{correct}) = P_e,$$
$$= P(\hat{C} \neq C)$$

be the probability of misclassification.

There are a number of different error rates and, while we do not consider all of them further, it is instructive to consider the differences between them:

- P_e^A, the apparent error rate;

- P_e, the true error rate for this classifier trained on the training set;

- $\mathbb{E}_D(P_e)$, the expected value of the error rate over training sets;

- $P_e^{\omega^*}$, the "least false" estimate with the "least false" parameters ω^*. This will happen when the classifier can not model the true decision surface;

- P_e^{Bayes}, the Bayes error rate. $P_e^{\omega^*} = P_e^{\text{Bayes}}$ if the classifier makes a correct distributional assumption. In the case of very flexible classifiers, such as the MLP, $P_e^{\omega^*}$ may be very close to P_e^{Bayes}.

The most direct measure is P_e^A, calculated as the proportion of errors in a data set. This is termed the *apparent error rate* or the *resubstitution error rate*. This is generally taken to be the error count $N - \delta\{C, \arg\max_c P(c|x)\}$, where C is the correct class and δ is the Kronecker delta, however in Section 5.2.2 (p. 56) a "smoothed" version of P_e^A is discussed.

The quantity that is really desired is an unbiased estimate of the expected error rate on unseen data, hence it is P_e that we are interested in knowing. However P_e^A is the only error rate directly available to us. Estimating P_e from P_e^A on the training data generally biases it downwards and gives overly optimistic results. Calculating P_e^A on an independent test set gives an unbiased estimate of P_e.

5.2.1   Re–sampled error rate

In many applications where classified data are very expensive to acquire, using an independent test set may be a very wasteful procedure. A number of alternative procedures, all based on resampling schemes, have been investigated. These could be broadly described as attempts to use the full set of training data to produce an estimate of P_e corrected for the bias of the apparent error rate. Hand (1997) identifies three variants of resampled error rate estimation:

1. the leave-one-out estimate;

2. the v-fold cross-validation; and

3. the bootstrap.

The leave-one-out estimate of the true error is

$$\frac{1}{N}\sum_{n=1}^{N}\rho^{(n)},$$

where the notation $\rho^{(n)}$ indicates that ρ is calculated leaving out the nth observation but then evaluated for the nth observation. It can be shown that this has a reduced bias compared to P_e^A, but there is evidence that this estimator has a relatively large variance compared to v-fold cross-validation (Ripley, 1996, §2.7).

For v-fold cross-validation the sample is divided into v parts (which we assume are equal in size) and the classifier is then trained on $\dfrac{v-1}{v}$ of the sample and tested on the excluded $1/v^{th}$ part. This is done, in turn, for each of the v parts and the average of the errors is taken.

The bootstrap attempts to estimate the bias $B = P_e - P_e^A$. The procedure is to sample, with replacement, from the training data, T. let T_k be the kth "bootstrap" sample and let $P_e^{A_{T_k}}$ be the apparent error rate on T_k. Let $P_e^{A_{T^k}}$ be the apparent error rate of the original data with the classifier developed on the T_k bootstrap sample. Then the kth bias estimate is $B_k = P_e^{A_{T_k}} - P_e^{A_{T^k}}$. We then estimate the bias as $B = 1/K \sum_{k=1}^{K} B_k$, where we have K samples from the training data. Then

$$P_e^{boot} = P_e^A + B.$$

When an observation from T is in T_k then the classifier trained on T_k may predict that observation well. In order to overcome this dependence on the presence or otherwise of particular observations, Efron (1983) suggested the "0.632" estimator as a variant on the bootstrap. Let $P_e^{A_{T^{\bar{k}}}}$ be the apparent error rate of those observations in the original data that do *not* appear in the kth bootstrap sample. Then let $B = 1/K \sum_{k=1}^{K} P_e^{A_{T^{\bar{k}}}}$. The "0.632" estimator is then

$$P_e^{0.632} = 0.368 P_e^A + 0.632 B,$$

where 0.632 is an approximation for $1 - (1/e)$.

Krzanowski and Hand (1997) suggest that in assessing the variability of the leave-one-out estimator, inappropriate criteria have been used. They argue that in using the Bayes error rate, P_e^{Bayes}, as the basis of comparisons rather than the true error rate, P_e, one has effectively averaged over different true error rates. However in any practical application one realization of the data has been obtained with one true (but unknown) error rate with a given type of classifier. It is for this error rate that we seek an unbiased estimator with small variance. However, after making this correction, the conclusion of Krzanowski and Hand (1997) is still that the leave-one-out estimator is poor and that the "0.632" estimator has superior qualities.

McLachlan (1992, §10) provides a discussion of cross-validation, the jackknife, the bootstrap and the "0.632" estimator. See also Hand (1997) for a discussion of the evaluation of classifiers. Note that for many of the models where leave–one–out resampling has been used successfully, dropping one observation from the training data and replacing it with another observation can be done by an "updating" operation that does not involve refitting the entire model. These methods may be unsuitable to the MLP model where the whole model has to be refitted.

5.2.2   A smoothed error rate

If $P(C_q|x)$ is available, it is possible to form a "smoothed" version of P_e^A that has a smaller variance.

$$P(\text{correct}|X = x) = P\{C = \arg\max_c P(c|x)|X = x\}$$
$$= \max_c P(c|x),$$

and so $1 - P_e = P(\text{correct}) = \mathbb{E}\{\max_c P(c|x)\}$. Thus $\max_c P(c|x)$ is an unbiased estimator of $P(\text{correct}|X = x)$ but, unlike the error count, $\delta\{C, \arg\max_c P(c|x)\}$, it is averaged over $P(C|X = x)$ and can be shown to have a smaller variance (McLachlan, 1992).

Writing the smoothed error estimate as P_e^S we have, following Ripley's (1996) argument,

$$P_e^S = 1 - \max_c P(c|x)$$
$$\leq 1 - \frac{1}{Q}.$$

Then

$$\mathbb{E}(P_e^S) = 1 - P(\text{correct}) = P_e$$

and

$$\mathbb{E}(P_e^{S^2}) \leq \left(1 - \frac{1}{Q}\right)\mathbb{E}(P_e^S)$$

so that

$$\mathbb{V}(\max_c P(c|x)) = \mathbb{V}(P_e^S)$$

$$\leq \left(1 - \frac{1}{Q}\right) P_e - P_e^2$$

$$= P_e - P_e^2 - \frac{P_e}{Q}$$

$$= \mathbb{V}\left[\delta\{C, \arg\max_c P(c|x)\}\right] - \frac{P_e}{Q}$$

If we have accurate estimates of $P(c|x)$ then this gives us a "smoothed" estimate of P_e^A, however Ripley warns that the effect of poor estimates of $P(c|x)$ can be severe. Note that we do not need to know the class C to calculate the smoothed error rate.

5.2.3   Further notes

In comparing two classifiers account should be taken of the fact that they are (generally) compared on the same data. If A and B are the two classifiers and n_A is the count of errors made by A but not by B, and likewise for n_B, then n_A has a $B(n_A + n_B, 1/2)$ distribution if the classifiers have the same error rate; this leads to McNemar's test (Dietterich, 1996) for testing equality of error rates.

In multi–class problems we may be interested in the pattern of errors. In this case we can give the "confusion matrix," with elements $\epsilon_{k_1,k_2} = \#(\text{classed as } k_2|\text{class } k_1)$. The term ϵ_{k_1,k_2}, divided by the number of observations gives the class conditional error rates for the training or test data. The misclassification measures considered above can be extended to the class–conditional error rates, including P_e^S (Ripley, 1995).

Ripley (1996) suggests using one of the resampling estimators in conjunction with P_e^S (thereby tackling both the bias and the variance problems). Venables and Ripley (1994) use a 10–fold cross-validation to give a resampled estimate of both the count error and the smoothed error P_e^S. In order to do this an MLP model is fitted and then allowed some re-training for each of the cross-validated data sets. This may be a source of some bias (see Section 9.4.3, p. 153).

5.3   MODEL SELECTION FOR MLP MODELS

We initially consider model selection for linear models before considering the MLP. Model selection for neural networks is generally accomplished by pruning, although there are also network growing and combined growing/pruning algorithms (Alpaydin, 1991). Reed (1993) gives a survey of pruning methods.

5.3.1   Linear models

For a linear model,

$$\underset{N \times 1}{y} = \underset{N \times PP \times 1}{X} \underset{}{\omega} + \underset{N \times 1}{\epsilon}. \tag{5.1}$$

with $\mathbb{E}[\epsilon] = 0$ and $\mathbb{E}[\epsilon\epsilon^T] = \sigma^2 I_N$, we derive $\hat{\omega} = (X^T X)^{-1} X^T y$. Now, if we have two models with design matrices $\underset{N \times P}{X}$ and $\underset{N \times P_0}{X_0}$ such that $R(X_0) \subset R(X)$, where $R(X)$ is the range or column space of the matrix X, then writing $\eta = X\omega$, we can test the hypothesis

$$H_0 : \quad \eta \in R(X_0)$$

versus

$$H_1 : \quad \eta \in R(X)$$

via the following. Define the deviance as

$$\mathbb{D} = 2 \left\{ \log f_{\max} - \log f(\hat{\omega}) \right\}, \tag{5.2}$$

where $\log f_{\max}$ is the log–likelihood of the maximum model and $\log f(\hat{\omega})$ is the log–likelihood of the current model. For (5.1), assuming Gaussian noise and a saturated maximum model, the deviance reduces to

$$\mathbb{D}(X) = (y - \hat{y})^T (y - \hat{y})$$

$$= y^T (I - X(X^T X)^{-1} X^T) y.$$

Then write

$$F = \frac{\dfrac{\mathbb{D}(X_0) - \mathbb{D}(X)}{P - P_0}}{\dfrac{\mathbb{D}(X)}{N - P}}.$$

If H_0 is true $F \sim F(P - P_0, N - P)$, which allows us to build a lattice of nested models and then select the most parsimonious model that describes the data.

5.3.2   The curse of dimensionality

The "curse of dimensionality" is the name given to the fact that the size, N, of a uniformly distributed training set must grow exponentially with P, the dimension of the feature space, in order to have the same average number of data points per unit volume of the feature space. This has serious implications for modeling data with a limited number of observations.

Bishop (1995a, §1.7) suggests that the MLP is not subject to the curse as ρ_l falls with order $O(1/H)$ and not $O(1/H^P)$ as would be typically expected[1]. Ripley (1996, §5.7) argues that this is not correct and that neither the MLP nor any similar non–linear method can break the curse of dimensionality. See Donoho (2000) for a wide ranging discussion of the "curse."

Hamamoto et al. (1996) report a simulation study on the generalization error when the ratio of sample size to dimensionality is small. The training data consisted of two classes drawn from Gaussian distributions with various separations between the means. There were P observations in each class, where P was the dimension of the feature space and H was fixed at either 8 or 256. It appears that for quadratic

---

[1]This ignores some constant factors. See Ripley (1996, proposition 5.3) for more detail.

discriminant functions, nearest neighbor, and Parzen window classifiers, the generalization error climbs with the dimensionality of the feature space, whereas it is essentially unchanged for the MLP classifier. This is not quite the same situation as above (where H is allowed to increase and the other factors are held constant) however it is interesting. Hamamoto et al. offer no explanation and none presents itself at the moment.

5.3.3 Computational learning theory and the degrees of freedom

The question of exactly what functions can be learned is considered by computation learning theory or "the theory of the learnable". Cybenko (1990) attempts to quantify the complexity of a learning task. Baum (1990) suggests that if we have N observations and an MLP with R weights that can learn all N observations, then we can expect $\epsilon = \frac{R}{N}$ errors in classifying future examples. In a similar vein Morgan and Bourland (1991) quote a rule of thumb that the number of observations should be 10 times the number of weights. See also Baum and Haussler (1989).

Ripley (1996, §2.8) reviews the area and describes the results of computational learning theory as very conservative, and indeed there seems ample evidence in the literature that MLPs with much greater ratios of weights to observations perform quite successfully. In fact, MLPs are frequently fitted with more weights than observations. Goodacre et al. (1992) give an example involving the detection of adulteration in olive oil that has $P.H.Q = 150.8.1$ and the MLP is trained on 11 observations. This would suggest that the degrees of freedom is negative and the models over–parameterized. Assuming that the inputs were of full–rank and the hidden–layer and activation functions were linear, there would be an infinite set of weights that would give a penalty function value of zero.

However the hidden layer and activation functions do make a difference. The concept of "degrees of freedom" is only relevant to linear models that are fitted by least squares. It is not uncommon in chemometrics, for example, to use more parameters than observations, even with linear models, but in this case the models are not fitted by least squares[2].

Attempts have been made to extend the concept of degrees of freedom to non-linear models and other forms of fitting, for example, Buja et al. (1987) extend the concept to linear smoothers, and Moody (1992) introduces p_{eff}, the "effective degrees of freedom." These measures of the degrees of freedom all tend to come out as less, often much less, than the number of weights in an MLP model.

In addition, as many MLP models are not trained to the global minimum of least squares, the model has not been fitted in the same sense that a linear model has been fitted. Hence counting parameters will very rarely be relevant as some of these parameters are not used effectively, and thus enter into the model in a harmless way. It is in this context, of ensuring that the excess parameters are of limited harm, that penalization techniques are used (as discussed in Section 5.4).

Bishop (1995a) shows that the effective number of parameters increases during training. Utans and Moody (1991) show that the effective number of parameters is a function of the regularization parameter and is decreased by regularization.

Bartlett (1998) shows that the generalization performance of an MLP depends more on the L^1 norm of the Ω weights, $\sum |\text{vec}(\Omega)|$ than on the number of hidden

---

[2]This point was made by B. Ripley in a posting to `comp.ai.neural-nets`.

layer units. Ingrassia and Morlini (2005) take this up and propose as a measure of the effective degrees of freedom, $p_{\text{eff}} = \text{trace}(P_\Omega)$, where $P_\Omega = \Omega(\Omega^T\Omega)^{-1}\Omega^T$ is the projection matrix onto the columns of Ω. They suggest using this with criteria such as Schwartz's Bayesian Information Criteria (Section 5.3.5). See also Chapter 13 (p. 219), where Bartlett's ideas are taken up in a different way.

While the degrees of freedom of an MLP is more difficult to quantify than in a linear model, there is no doubt that an excessive number of hidden layer units plus a long period of training can lead to overfitting. See Chauvin (1990), Ahmed and Tesauro (1988), and Morgan and Bourland (1991) and many others on this question. See also experiment 3, Section 9.3, p. 145.

5.3.4   Wald tests

There are several pruning strategies based on approximations to the Wald test (Silvey, 1975, §7) which consider the ratio of a weight to its standard error, calculated as the appropriate diagonal of the inverse of the Fisher information matrix

$$\mathcal{I} = \mathbb{V}\left[\frac{\partial\rho}{\partial\omega}\right],$$

where ρ is a negative log–likelihood for some assumed parametric family f. The test statistic[3] is a χ^2 statistic of the form $\omega^T\mathcal{I}\omega \sim \chi^2(R)$ if the hypothesis, $\omega = 0$, is true.

The practice in the MLP literature is to calculate the statistic $\omega_r^2/\mathcal{I}_{rr}^{-1}$ for all of the weights and delete the weight with the smallest statistic value. The methods differ in how $\mathcal{I}$ is approximated. Le Cun et al. (1990) use

$$\mathcal{I} \approx \text{diag}\left(\frac{\partial^2\rho}{\partial\omega_r\partial\omega_r}\right),$$

the diagonal of the observed information. However the assumption that $\mathcal{I}$ is diagonal is often a poor one (Bishop, 1995a, §9.5). Hassibi and Stork (1991) use the covariance matrix of the scores $\partial\rho/\partial\omega$ so

$$\mathcal{I} \approx \left(\frac{\partial\rho}{\partial\omega}\right)\left(\frac{\partial\rho}{\partial\omega}\right)^T,$$

and then calculate the inverse. See Ripley (1996, §5.6) and Bishop (1995a, §9) for discussion.

White (1989) suggests a related test for redundant units based on the result that

$$\sqrt{N}(\omega_N - \omega^*) \xrightarrow{D} N(0, C), \tag{5.3}$$

where

$$C = A^{-1}B\mathcal{I}^{-1}$$

$$A = \mathbb{E}\left[\frac{\partial^2\rho}{\partial\omega^2}\right], \tag{5.4}$$

$$B = \mathbb{V}\left[\left(\frac{\partial\rho}{\partial\omega}\right)\right],$$

---

[3]R is the total number of weights. See Section 2.1, page 12.

The ω_N are the weights for a sample of size N, and ω^* are the optimal weights. This is a multivariate t-test of the form

$$H_0 \quad S\omega^* = 0 \text{ versus } H_1 \quad S\omega^* \neq 0,$$

where S is a matrix that selects the weights associated with the hidden units hypothesized to be 0. Equation (5.3) is an extension of the standard result (Silvey, 1975) that

$$\sqrt{N}(\hat{\omega}_N - \omega^*) \xrightarrow{D} N(0, \mathcal{I}^{-1}) \tag{5.5}$$

to the case where the true density f^* is not a member of the family f. $\hat{\omega}$ is then the "least false" estimate arising from a misspecified model in the sense that the Kullback–Liebler divergence

$$d(f^*, f(\hat{\omega})) = \int f^* \log\left(\frac{f^*}{f(\hat{\omega})}\right) dx$$

is minimized over the parametric family f (Ripley, 1996, §2.2). However, if $f^* \in f$ then $A = B = \mathcal{I}$ and (5.3) collapses to (5.5).

In addition to the usual problem of asymptotic results (i.e., is the convergence fast enough to be usable in realistic examples?) there is the problem that the information matrix of an MLP may be singular and equation (5.3) invalidated if certain conditions are not met. These conditions include that the network be "irreducible" and that there be no collinearities in the features. An irreducible network is one such that there is no smaller network that gives the same input–output mapping (Fukumizu, 1996). Hence (5.3) will not apply in the very circumstance that it is hoped to test for, namely the presence of redundant units.

In addition to the above conditions, this test will fail in conditions where the weights associated with a redundant unit are not of small magnitude[4]. See also Liestol et al. (1994), who urge caution in the use of such tests for removing hidden-layer units.

5.3.5   Network information criteria

The Network Information Criterion (NIC) (Murata et al., 1994) is a penalty approach based on the precept that the deviance will be greater for a test set than for the training set, and an attempt to quantify this difference. Closely related (and antecedent) approaches are Akaike's Information Criteria (AIC), Mallow's C_p, Schwartz's Bayesian Information Criteria and Barron's Predicted Squared Error (Barron, 1984). If we write $\mathbb{E}(D_1)$ as the expected deviance for a single test example then

$$N\mathbb{E}(D_1) = \mathbb{E}(\mathbb{D}) + 2R^* + O\left(\frac{1}{\sqrt{N}}\right), \tag{5.6}$$

where R^* is a complexity measure defined below. Ignoring the $O(\frac{1}{\sqrt{N}})$ term and substituting $\mathbb{D}$ for $\mathbb{E}(\mathbb{D})$, we derive the NIC, $\mathbb{D} + 2R^*$, and we can fit a hierarchy of nested models and select the model for which the NIC is a minimum. Note that the models have to be nested otherwise the final term can not be guaranteed to

---

[4]There is an example of this in experiment 2 case 2, Section 9.3, p. 145.

be of order $\frac{1}{\sqrt{N}}$ (Murata et al., 1994). Equation (5.6) is also applicable when $\mathbb{D}$ is replaced by a penalized deviance, as discussed in the Section 5.4, p. 62.

If the parametric family contains f^* then $R^* = R$, the number of parameters, and (5.6) reduces to the AIC, else $R^* = \text{trace}(BA^{-1})$, where A and B are as defined in (5.4). Moody's 1992 "effective degrees of freedom" p_{eff} gives an equivalent form for R^*, see also Moody and Utans (1995) for a comparison of cross–validation and NIC model selection.

Ripley (1996, §5.6) cautions that both p_{eff} and NIC depend on the penalty function surface having a single pronounced local minimum. If this condition is not met then the effective degrees of freedom may depend more on the local minimum found than the number of hidden layer units. See also Ripley (1996) §2.2 and §4.3 for a derivation and treatment of the NIC.

5.3.6 Other methods

Sietsma and Dow (1991) suggest interactive pruning guided by heuristics. Utans and Moody (1991) use generalized cross–validation and Squared Prediction Error (Barron, 1984) to do model selection, but use the number of parameters as the degrees of freedom. Mozer and Smolensky (1989) suggest pruning units on a measure of "relevance" that is an estimate of $\rho_{(r)} - \rho$ where $\rho_{(r)}$ is the penalty function with unit r removed. See also Chapter 6 (p. 69), where task–based pruning is introduced and discussed. This method relies on testing for similarities in the outputs of hidden layer units.

5.4 PENALIZED TRAINING

Penalization appears in linear models under the name "ridge regression". If the matrix $X^T X$ is ill conditioned[5] then, writing $\lambda_{\min}$ as the smallest eigenvalue of $X^T X$ and ω^* as the true value, $\dfrac{\sigma^2}{\lambda_{\min}}$ is a lower bound for the average squared distance between $\hat{\omega}$ and ω^* (see Marquardt, 1970, for a discussion), and thus the estimates may be quite poor. In these circumstances Marquardt comments that the absolute values of the ωs may be too large and may change sign with negligible changes in the data. A better procedure in these circumstances is to solve $\hat{\omega} = (X^T X + kI)^{-1} X^T y$ for some constant k. This inflates the eigenvalues of the $X^T X$ matrix, giving a biased estimator with perhaps a lower expected mean square error.

In order to rectify the tendency of MLPs to overfit the data, a number of similar penalization methods have been tried. All are methods for obtaining a smoother separating boundary between classes, and all involve some trade–off between error minimization and smoothness via the selection of a smoothing parameter. These methods have all been introduced in the context of decreasing the error of the MLP

---

[5]See Stewart (1987) for some history of the concept of matrix condition. It appears to have developed separately within the statistics and numerical analysis communities. Within numerical analysis it can be traced through Turing (1948) and Golub and Wilkinson (1966). Piegorsch and Casella (1989) give a history of matrix conditioning and ridge regression in statistics. The condition number $\kappa(A) = \|A\|\|A^{-1}\|$ is a measure of the stability or sensitivity of a matrix (or the linear system it represents) to numerical operations. If it is large then numerical operations involving the matrix may have a large error.

on previously unseen data by restricting the MLP in the way that it fits the training data.

The methods include the following:

1. Adding noise to the training data. Sarle (1995) coins the term "smoothed ridging" for this method and Bishop (1995b) shows that it is approximately equivalent to the Tikhonov regularization term

$$\frac{1}{2N}\sum_n\sum_p\sum_q\left(\frac{\partial z_{nq}^*}{\partial x_{np}}\right)^2 \tag{5.7}$$

when the amplitude of the noise is small. See Sietsma and Dow (1991) for an example;

2. Scaling the weights down after training;

3. Replacing the target value with the convolution of the target and a probability distribution;

4. Adding a regularization term to the error function. This is "penalized ridging" (Sarle, 1995) and can be justified within a Bayesian framework. It appears in the literature in two forms. One form is the case where the penalty term is of the form

$$\sum_q\frac{1}{2}(z_q^* - t_q)^2 + \lambda||Dz^*||^2, \tag{5.8}$$

where D is a differential operator. Equation (5.7) is a finite sample version of such a differential operator regularization term. Note that (5.7) and (5.8) involve derivatives of z^* with respect to x. In order to fit these penalized models the derivatives with respect to the ω are needed, see Bishop (1993) for the computations. Ripley (1996, §4.3) notes that as regularization has been most explored in approximation theory, the results are generally for least squares approximation. Poggio and Girosi (1990a,b) and Girosi et al. (1993) discuss regularization in the context of surface reconstruction from sparse data points. The approach is motivated by the recognition that the problem of function approximation from sparse data is ill–posed in the sense that the solution is not unique, and that additional constraints are needed in the form of an appropriate prior on the class of approximation functions to guarantee the uniqueness of the solution. Often the prior is one that assumes a high degree of smoothness. They show that (5.8) leads to a class of hidden–layer networks that they call "regularization networks" of which radial basis function networks are a subset. If G is the Green's function (Courant and Hilbert, 1953) of P^*P where P^* is the adjoint of P then (5.8) leads to

$$z^*(x) = \sum_{n=1}^{N}\omega_n G(||x - x_n||)$$

and gives a fitted surface equivalent to a generalized spline. In particular if P is $\int\partial^2 z^*/\partial x^2 dx$ then (5.8) leads to the cubic spline.

5. An alternative form of regularization where the prior represents *a priori* knowledge other than smoothness of the fitted surface. In the case of the

MLP it may be a prior belief that many of the weights are redundant and could be set to 0. This leads to the prior $\omega \sim N(0, \lambda I)$ and the penalty term

$$\rho(z^*, t) + \lambda \sum_r \omega_r^2.$$

This penalizes large weights and is known as weight decay (Plaut et al., 1986). It reduces to ridge regression for an MLP with no hidden layer and linear output functions. Williams (1995) and Goutte and Hansen (1997) suggest that the use of a Laplacian prior $\sum_r |\omega_r|$ has a number of desirable properties. This is related to the LASSO regression penalty term described in Tibshirani (1996);

6. Introducing a "weight elimination" penalty term (Weigend et al., 1991)

$$\rho(z^*, t) + \lambda \sum_r \frac{\omega_r^2}{1 + \omega_r^2}.$$

This is approximately equivalent to an improper prior of a Gaussian contaminated by a uniform distribution and penalizes large weights less than small weights, causing small weights to go to 0. It is essentially a variable selection technique. Weigend et al. (1991) also consider allowing the λ term to vary during training;

7. Using a penalty function consisting of a mixture of Gaussians (Nowlan and Hinton, 1992), i.e.,

$$\rho(z^*, t) = \frac{1}{\sigma_{z^*}^2} \sum_q \frac{1}{2}(z_q^* - t_q)^2 - \sum_r \log \left(\sum_j \pi_j p_j(\omega_r) \right),$$

where p_j is a Gaussian with mean μ_j and variance σ_j^2. Then μ_j, σ_j^2, π_j and $\sigma_{z^*}^2$ are all estimated as parameters in the model. The thinking behind this prior is that there are likely to be groups of similar weights, some perhaps large and some small. Hence, as with weight decay, if some of the weights can be seen to be grouped around zero, these weights can be discarded. The approach is called "soft weight sharing".

8. Stopping training early. This involves dividing the data into a training and a test set and training on the training set until the penalty function starts to rise on the test set. As shown in Bishop (1995a, §9, see also Exercise 9.1) the degrees of freedom increase during the training process. Hence by stopping early, one effectively fits a model with fewer degrees of freedom. See also Section 9.4.3 (p. 153), where it is argued that the changing shape of the IC during the fitting process may cause the MLP to become progressively less resistant to some features of the training set. Moody and Utans (1995) suggest that stopped training only becomes practical when the data sets are very large, else the training and test sets do not contain sufficient information. However, Amari et al. (1997) give contrary advice. They suggest that for large enough samples (larger that $30R$) training on the full training set is better than stopped training. In addition they give a formula for determining the optimal proportion of the available data to use for training and test sets.

We note that when a test set is used in this way, it is no longer independent and does not give an unbiased estimate of the error rate. Many practitioners call an independent, unused, test set a "validation set" to distinguish it from an early stopping test set.

Ripley (1996, §5.3) warns that the procedure may be harder to apply in practice as the test-set error may pass through a minimum and rise for a large number of iterations only to fall again to a new lower local minimum.

Reed et al. (1995) describe asymptotic situations under which methods 1, 2, 3 and 4 (with $P = \dfrac{\partial z^{*}}{\partial x}$) give similar results. Sarle (1995) gives an empirical study of 4 and 8 and a discussion of some issues in Bayesian estimation. Weigend (1994) suggests that a large model used with regularization may give an improved performance over a smaller model. In addition it is suggested that even a small model may overfit the data in the sense of Weigend's operational definition of overfitting. Morgan and Bourland (1991), using stopped training, report that over–parameterization leads to poor generalization, but as their training and test sets are correlated[6], they may not have been stopping training early enough. See also Chapter 13 where fixed-slope training is introduced and discussed.

Venables and Ripley (1999, §11.4) describe a crude but effective approximation to Bayesian training. We note such approximations are quite desirable in this context and an attempt to do Monte Carlo integration over the weight space (for example in Neal, 1995) is likely to lead to training times that are large even by the standards of the neural network community.

5.5   COMPLEMENTS AND EXERCISES

5.1   The `mlp` library provides the weight decay penalty term (item 5, p. 64),

```
mlp.model1<-mlp(n,p,h,q,data,target,lambda1=0.001)
```

and the weight elimination penalty term (item 6, p. 64),

```
mlp.model1<-mlp(n,p,h,q,data,target,lambda2=0.001).
```

In Figures 5.1, 5.2 and 5.3 we see the effect of these penalization terms. Figure 5.1 shows the result of fitting a standard MLP of size 2.3.1. Figure 5.2 shows the same model trained with weight decay ($\lambda = 0.001$). As the magnitude of the weights is restrained the MLP does not form a decision region with sharp corners and this tends to make the model more resistant to individual points that would require an irregularly shaped decision region to accommodate them. Figure 5.3 shows the same model trained with weight elimination. This acts like a variable selection technique and causes two of the hyperplanes to be orthogonal to the axes.

Using the supplied code, refit these models and examine the weights. Note that the weight elimination model eliminates weights, not nodes.

5.2   This is a relatively easy problem due to the fact that there is a good separation between the two classes. To make the problem more difficult, try:

[6]This is pointed out in Sarle (1995).

- moving one of the dark class over toward the light class, and then inside of the boundary of the light class;

- adding more hidden layer units to the MLP.

It is likely that you will see the standard MLP decision boundary tracking the moving point and perhaps making a small region around it. The penalized MLP models may behave differently.

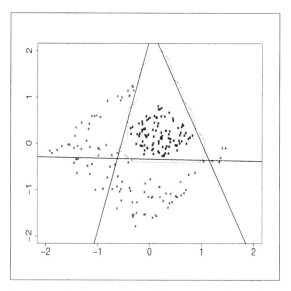

Figure 5.1   A standard MLP model of size 2.3.1. The lines formed by the three hidden layer units are shown, as is the final triangular shaped decision region.

5.3   Show that for a linear regression the degrees of freedom are equal to the trace of the "hat" matrix $X(X^T X)^{-1}X^T$. Write down the "hat" matrix for a ridge regression and show (by a numerical example or otherwise) that the effective degrees of freedom are reduced as the λ penalty coefficient is increased.

For linear smoothers, $\hat{y} = Sy$, where S is the smoothing matrix, the trace of S is often used as a measure of effective degrees of freedom.

5.4   By making a quadratic approximation to $\rho_l + \lambda \sum_r \omega_r^2$ at the solution, write down an expression equivalent to the smoothing expression, $\hat{y} = Sy$, in terms of the Hessian matrix. Hence write down an expression form the effective degrees of freedom in terms of the eigenvalues of the Hessian (Bishop, 1995a).

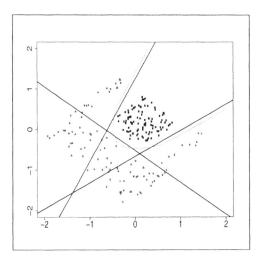

Figure 5.2   This figure shows the same model as Figure 5.1, trained with weight decay ($\lambda = 0.001$). As the magnitude of the weights is restrained the MLP does not form a decision region with sharp corners. This tends to make the model more resistant to individual points that would require an irregularly shaped decision region to accommodate them.

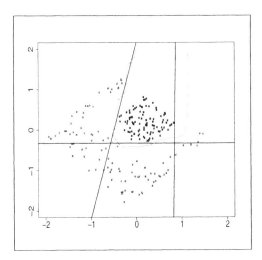

Figure 5.3   The figure shows the same model as Figure 5.1, trained with weight elimination. This acts as a variable selection technique and causes two of the hyperplanes to be orthogonal to the axes.

CHAPTER 6

THE TASK–BASED MLP

6.1 INTRODUCTION

We examine the question of the MLP topology and a pruning methodology is considered. The method is based on a design strategy arising from the fact that, in the case of a classification network with one hidden layer, it is possible to select an architecture that ensures there is at least one separating hyperplane for each pair of classes. Building on this idea, the "task–based MLP" is developed[1].

The actions of the hidden-layer units can be understood in terms of separating classes or super-classes and the Υ weights as combining these decisions to form the final classifications. Furthermore the actions of the Υ weights can be readily interpreted for a task-based MLP and that this interpretation may lead to ways of considerably simplifying the final model.

Based on the task–based MLP, a pruning strategy is developed and a practical application to multi–spectral image data is presented. It is shown, for this example, that a large MLP (of size 6.15.6) can be considerably pruned (to 6.4.6), and that, in addition, many of the entries in the Υ matrix can be set to integer values (including many zeros) without loss of classification accuracy. In addition a graphical method is developed for tracking the pruning schedule.

---

[1]It was first introduced in Dunne et al. (1992), and has similarities to the model proposed by Knerr et al. (1990).

A Statistical Approach to Neural Networks for Pattern Recognition by Robert A. Dunne   **69**
Copyright © 2007 John Wiley & Sons, Inc.

6.2   THE TASK–BASED MLP

Prior knowledge is incorporated into an MLP via the starting weights and the architecture. Hence the practice of starting an MLP with an arbitrarily chosen architecture and random weights to learn a classification task is tantamount to taking the position that there is no knowledge about the position of the decision boundaries. However, in an obvious way, this does not accurately reflect the available prior knowledge. For example, the separating hyperplanes must at least intersect the region of the feature space occupied by the classes. To incorporate this knowledge, we could select the initial values of the weights of the MLP so that the resulting hyperplanes intersect the region of the feature space occupied by the classes.

We start from the standard MLP model, of size $P.H.Q$, described in Section 2.1 (p. 9). For the task–based MLP, we assign each hidden–layer unit a "task" which is to separate a particular pair of classes. Hence H will be equal to $\binom{Q}{2}$, the number of pairs of classes.

There are a number of ways of approaching this. One way is to select the initial weights randomly and then train each hidden–layer unit only on the classes that it is designed to separate. In this case, the network is similar to the one introduced by Knerr et al. (1990) and briefly described in Section 6.2.1. In addition to ensuring that the hidden–layer unit does its ascribed task, this would ensure that the outputs of the hidden–layer unit are < 0.5 for typical examples of one class and > 0.5 for examples of the other class, which is necessary for the pruning algorithm described in Section 6.3.3, p. 74.

However, the approach adopted here is to calculate Fisher's linear discriminant function (Section 3, p. 19) for each pair of classes. The LDF gives a normal, v, to a hyperplane chosen to maximize the separation of the classes when they are projected onto it. We take the hyperplane given by v and passing through $\bar{x}_g$, the grand mean of the classes. Thus we set the initial estimate of the corresponding row of Ω to be $\left\{ -(\bar{x}_g^T v), v^T \right\}^T$

6.2.1   Single–layer learning

Knerr et al. (1990) present an incrementally constructed classification network, the "single layer-learning network," designed to overcome the problem that the architecture of an MLP must be specified in advance (see also Knerr et al. (1992) for an application to handwriting recognition). This network results in a similar hidden–layer to the task–based MLP.

For the single–layer network, each unit in a layer starts with random weights but, unlike a standard MLP, each unit responds only to exemplars from two classes. Hence, after training, the network should have positioned a separating hyperplane between each pair of classes.

A difference between the single–layer architecture and the task–based network is that the single–layer learning network first of all tries to separate each class from all the other classes combined. Thus, if there are Q classes, then Q units are used to attempt the separation. If m are successful, the other $Q - m$ are discarded and an all–pairs separation is attempted for the $Q - m$ remaining classes. This may lead to a smaller hidden layer than the task–based network for some problems.

An example of the decision surfaces arrived at by a single–layer learning network is illustrated in Figure 6.1, which shows a classification problem involving five classes in two dimensions. Class 1 is separable from all of the others combined and so would be isolated in the first stage of the algorithm. Pair–wise separation would then be attempted on classes {2,3,4,5}, needing 6 units. With the exception of classes 4 and 5, these classes would be found to be linearly separable. Below the unit assigned to separate classes 4 and 5, a sub–network would be implemented.

An inspection of this diagram shows that three of the units are redundant from the point of view of positioning a separating hyperplane between each pair of classes, and so this "single–layer" network has an excess number of hidden–layer units.

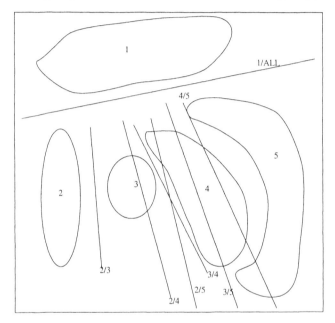

Figure 6.1   An example of the decision surfaces arrived at by a single–layer learning network is illustrated for a classification problem involving five classes in two dimensions. Class 1 is separable from all of the others combined and so would be isolated in the first stage of the algorithm. Pair–wise separation would then be attempted on classes {2, 3, 4, 5}, using 6 units. With the exception of classes 4 and 5, these classes are linearly separable. Below the unit assigned to separate classes 4 and 5, a sub–network would be implemented. After Knerr et al. (1990), p. 42.

6.3   PRUNING ALGORITHMS

The term "pruning" is taken from its use in classification trees (Breiman et al., 1984, §10) and is used here to indicate that we are removing unwanted structure in order to leave a more parsimonious model. See Section 5.3 (p. 57) for a brief discussion of the literature on pruning MLP models.

A problem with both the single–layer and the task–based approaches is that the resulting hidden layer may be of large size. For the task–based MLP, the hidden layer grows at $O(Q^2)$ where Q is the number of classes. The nature of this layer suggests some pruning ideas which are illustrated by way of the following simple example. Consider four classes, each at the corner of a square in a two–dimensional feature space (Figure 6.2). This will need six discriminant functions for a full pairwise separation, and these are also illustrated. Note that while each unit has been selected to perform one task of separating a particular pair of classes, it may also separate other pairs of classes. A subset of units, that performs essentially the same tasks as the full set, should give as good a separation of the data as the full set of units.

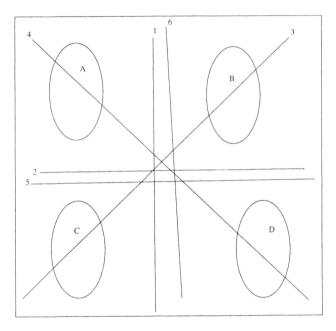

Figure 6.2   Four classes in a two–dimensional feature space and a representation of the six linear discriminant functions to pairwise–separate them. The discriminant functions have been given an arbitrary numbering which is used to refer to them in the following argument.

Using the hidden–layer output, we can calculate a diagnostic matrix T, of size $Q \times H$, where T_{qh} is the number of "+" results $(f_1(\omega^T x) > 0.5$ so that $\omega^T x > 0)$ for the qth class given by the hth hidden–layer unit. An inspection of this matrix, given in Table 6.1, indicates which classes are being separated by the hyperplanes in feature space. An analysis of this matrix can deduce which tasks are being performed by which unit or discriminant function. This information is summarized in Table 6.2, which gives the task performed by each unit, an arbitrary numbering of that task, and a list of other tasks performed by that same unit. It is clear that many of the hidden–layer units duplicate tasks assigned to other units.

As the matrix $f(X\Omega^T)$ (and the matrix T, which is a compressed version of $f(X\Omega^T)$) describe the separation achieved by the hyperplanes, an analysis of these matrices should indicate which units can be removed. We now consider three different techniques that could be applied to this task:

1. the QR decomposition;

2. hierarchical cluster analysis; and

3. task–based pruning.

Table 6.1   The T matrix for the example illustrated in Figure 6.2. T_{kj} is the number of "+" results for the kth class given by the jth hidden–layer unit. The number of the discriminant function is shown at the top of the table and the class in the first column; n_q is the number in class q. From this matrix, a simple analysis can deduce which tasks are being performed by which unit or discriminant function. This information is summarized in Table 6.2, which gives the task performed by each unit.

	1	2	3	4	5	6
a	n_a	n_a	n_a	$n_a/2$	n_a	n_a
b	0	n_b	$n_b/2$	0	n_b	0
c	n_c	0	$n_c/2$	n_c	0	n_c
d	0	0	0	$n_d/2$	0	0

Table 6.2   A listing of the discriminant functions, with the same numbering as in Figure 6.2, together with (column 2) the classes which the discriminant function is designed to separate, and (column 3) the numbers of any discriminant functions it performs the same task as.

discriminant function number	which was designed to separate classes	also does the same task as
1	a,b	1 3 4 6
2	a,c	2 3 4 5
3	a,d	3
4	b,c	4
5	b,d	2 3 4 5
6	c,d	1 3 4 6

6.3.1   The QR decomposition

The first technique is to calculate the QR decomposition (Noble and Daniel, 1988, §5.9) of the matrix $f(X\Omega^T)$. The absolute value of the diagonal elements of the R matrix give the distance of the hth column of $f(X\Omega^T)$ from the subspace spanned by the $1,\ldots,h-1$ columns (see Problem 6.2, p. 90). These distances can be normalized by dividing by the norm of the hth column of $f(X\Omega^T)$ and forming the vector

$$R_{hh}/\|f(X\Omega^T)_h\|_2.$$

Now if hidden–layer unit h is doing a different task to units $1, \ldots, h-1$, its normalized distance will be large. In the case where it does the same set of tasks, it will be in the subspace spanned by the $1, \ldots, h-1$ columns and its distance will be zero. Hence we can select the discriminant functions that are performing significantly different tasks.

In order to determine the dimensionality of $f(X\Omega^T)$, we plot the eigenvalues of $\{f(X\Omega^T)\}^T f(X\Omega^T)$. An inspection of this plot may indicate how many units should be pruned. In this case we will be hoping to see a sharp drop in the eigenvalues, rather than a continuous gradual decay.

6.3.2 Hierarchical cluster analysis

Another technique that could be applied is a hierarchical cluster analysis (Sneath and Sokal, 1973) of the columns of the $f(X\Omega^T)$ matrix. After we have clustered the units (using some distance measure), we can select one hidden–layer unit from each cluster and prune the others. This should produce a subset of the units that essentially performs the same task as the full set of hidden–layer units.

6.3.3 Task–based pruning

Clustering, QR decomposition and task–based pruning all rely on finding similarities between the outputs of hidden layer units. However, the similarity measures defined by clustering and QR decomposition are not sufficient to uncover all the similarities in the actions of hidden layer units. It is possible for the outputs of two units to be very dissimilar in these measures, and yet for the units to have a number of tasks in common.

Consider column 3 of Table 6.2 (p. 73). The proposed procedure is to start with the unit that performs the fewest tasks (ties are resolved by a random choice), and prune that unit if all of the tasks it performs are performed by other units. Repeating the sequence will result in a minimal set that performs all of the tasks. For example, we could start with unit 3, see that the task of separating a and d is also achieved by unit 1, and therefore prune 3. Next we would take unit 4 and also prune it. Then we might take unit 5, see that unit 2 performs all of 5's tasks and therefore prune 5, and so on. In this way we would end up with a minimal set such as units 1 and 2. Note that this minimal set is not unique and depends on the order in which we consider the units. If we had used a different order, we might have decided on the minimal set $\{5,6\}$. Clearly either of these sets will completely separate the four classes.

In this instance, the set $\{1,2\}$ will separate the data as well as the full set of six possible functions. In many instances, it may be desirable to allow a unit to be considered as performing a task if it does so with an error rate less than some given constant. If we are prepared to do this, we may be able to further prune the set of units and still end up with a set that will perform well in the testing phase.

The details of the algorithm are as follows. We use the T matrix to form two ragged arrays:

units.by.tasks which lists, for each unit, all tasks that the unit is performing. In this instance, performing a task means performing it as well as the unit that does the best at that task plus a preset tolerance;

tasks.by.units which lists, for each task, the units performing that task.

We then count the number of entries in the rows of **units.by.tasks** so that we know how many tasks each unit performs. The pruning algorithm is then:

1. pick, from the units not yet examined, the unit which performs the fewest tasks (breaking ties randomly) – say this is unit j;

2. check the matrix **units.by.tasks** to see what tasks j performs, and, for each task that j performs, check **tasks.by.units** to see if that task is performed by some other unit;

3. if all of j's tasks are performed by other units, then prune j;

4. if all units have been considered, then exit, else go to step 1.

6.3.4 Diagnostics

A number of diagnostic indicators can be used to monitor the pruning process. The first diagnostic measure is to ensure that all tasks have been performed. This is done by a direct check in the case of a full task–based MLP, where it is an easy matter to indicate which task each unit should be performing. In the case of a pruned MLP it can be done by calculating the classification matrix C, where C_{q_1,q_2} is the number of elements of class q_1 that are classified as belonging to class q_2.

In addition, we define the "confusion" of a hidden–layer unit to be the number of elements of the various classes for which it returns a minority result. That is, if for class q it returns 4 "−" results and $n_q - 4$ "+" results where $n_q - 4 \geq 4$, then 4 is added to the confusion score for that unit. The vector of confusion values for this example is given in Table 6.3 where we assume, without loss of generality, that the n_k are all even and that units 3 and 4 bisect the four groups. Unit 1 has 0 confusion, as it does not intersect any of the classes, whereas unit 3, for example, bisects classes b and c so its confusion is $n_b/2 + n_c/2$.

The classification matrix and the vector of confusion values can both be tracked throughout the pruning process and may provide a warning should problems develop with the pruning process.

Table 6.3   The confusion vector for the example illustrated in Figure 6.2.

1	0
2	0
3	$n_b/2 + n_c/2$
4	$n_a/2 + n_d/2$
5	0
6	0

6.3.5 Results

Trials of the three techniques revealed that task–based pruning performed considerably better than the other two proposed methods. In the remote sensing example considered in Section 6.4 below, QR pruning reduced the size of the hidden layer from 15 to 14 units while task-based pruning reduced it from 15 to 6 units. For a hierarchical cluster analysis the usual problems emerge:

- what is the appropriate distance metric?

- what is the appropriate cluster aggregation algorithm?

- where should we cut the dendogram?

With a complete linkage clustering algorithm, an Euclidean distance matrix, and a cut on the dendogram to retain 6 clusters, for a direct comparison with the task-based pruning algorithm, it was found that the six selected hidden-layer units corresponding to the clusters left a total of five tasks undone. It appears that the most direct method, task-based pruning, is the most effective.

6.3.6   Non–linear decision boundaries

So far we have assumed that the classes are linearly separable or that, because of distributional assumptions (Gaussian distribution with equal covariance matrices) or limited data, we have decided to restrict the decision boundaries to hyperplanes. If this restriction is not reasonable, then extensions to the algorithm will be needed.

Knerr et al. (1990) propose a method for dealing with the case where a non–linear decision boundary is appropriate. Consider two classes that are not linearly separable. After the decision hyperplane is fitted, if the classes are not separated, then the observations on one side of the decision hyperplane are passed on for consideration to another single–layer perceptron. This algorithm is similar to the "neural tree network" algorithm of Sankar and Mammone (1990a,b, 1991a,b) which involves placing a perceptron at each node on a decision tree. See also Sethi (1990) for a discussion of a hybrid decision tree – MLP algorithm and see Ripley (1996, §7.6) for an overview. A similar extension is possible for the task–based MLP.

6.4   INTERPRETING AND EVALUATING TASK–BASED MLP MODELS APPLIED TO MULTISPECTRAL IMAGE DATA (MARTIN'S FARM)

Remotely sensed data typically consist of a vector of measured reflectances at various wavelengths for each picture element (pixel) of an image. Such data may be gathered from many sources such as satellite and airborne scanners.

The data may then be used to create an image such as Figure 11.2 (p. 182). This is generally done in "false color" in order to get as much relevant information as possible into the 0–255 range of the 3 colors. Practitioners become adept at interpreting images in which vegetation appears as red etc.

Alternatively, a map is produced by assigning each pixel to a class of interest on the basis of its spectral signature. In common applications, these classes include such things as:

- different vegetation classes (crop versus bush);

- clouds versus snow and ice in polar images;

- sea grass versus turbid water.

To do this first requires the definition of training groups, which are typically homogeneous regions selected from the image by trained practitioners, using any available ground truth[2].

We consider the application of the task–based MLP to a classification problem involving such a multispectral image. Questions considered here are the optimality of the proposed pruning method and the simplification and interpretation of the resulting models.

The data are from a Thematic Mapper (TM) image of 301 pixels by 401 lines, with 6 bands for each pixel. The TM scanner is carried on the Landsat 4 and 5 satellites and returns seven bands, of which six are primarily used for vegetation monitoring; the 7th measures the thermal response at a lower resolution (120m × 120m pixel size). The bands and the wavelengths are listed in Table 6.4.

Table 6.4 The wavelengths of the seven bands returned by the Landsat Thematic Mapper scanner.

band number	wavelength (μm)
1	0.45 – 0.52 (visible blue)
2	0.52 – 0.60 (visible green)
3	0.63 – 0.69 (visible red)
4	0.76 – 0.90 (near infrared)
5	1.55 – 1.75 (middle infrared)
6	10.40 – 12.50 (thermal infrared)
7	2.08 – 2.35 (middle infrared)

The image was captured in August 1990 and includes a property known as Martin's farm, for which some ground truth is available. Figure 6.3 shows a ground truth map while Table 6.5 gives an explanation of the codes for the ground–cover classes. Using the ground–truth map, six rectangular training areas were selected. The six selected areas, each of 121 pixels, are described in Table 6.6, where the H designations refer to numbered fields on the ground–truth map. A grey–level image of the region (band 4) showing the six training areas is given in Figure 6.4.

It should be borne in mind that, due to soil and terrain differences and different planting dates, it may not be the case that the spectral characteristics of different fields of wheat, for example, are drawn from identical distributions. Because of this, it is desirable to restrict the decision boundary to hyperplanes so as not to over–fit the training data.

---

[2]Ground truth is information obtained by direct measurement at ground level, rather than by interpretation of remotely obtained data, especially as used to verify or calibrate remotely obtained data. Its older meaning is that of a "fundamental truth" and predates remote sensing.

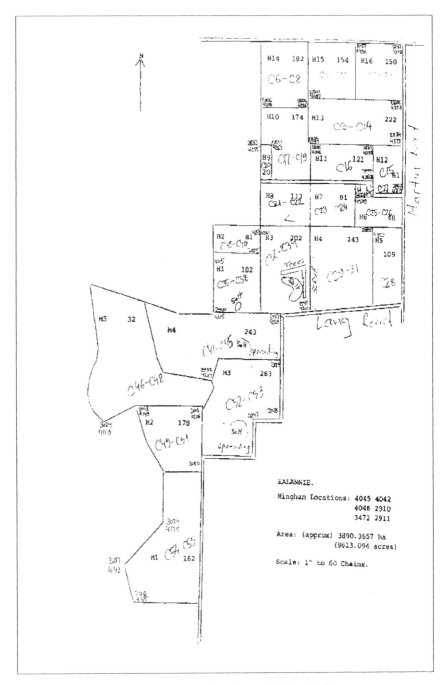

Figure 6.3   The ground–truth map for Martin's farm.

Table 6.5   The ground–cover code used in the ground–truth map of Martin's farm in Figure 6.3. M. pasture and V. pasture are two different varieties of pasture crop. H3 and M3 were planted with two different crops.

field identifier	ground cover	notes
H1	lupins	salt
H2	wheat	
H3	wheat and lupins	salt
H4	wheat	salt
H5	unknown	
H6	V. pasture	
H7	wheat	
H8	wheat	
H9	lupins	
H10	V. pasture	
H11	V. pasture	
H12	wheat	
H13	V. pasture	
H14	wheat	
H15	V. pasture	
H16	V. pasture	
M1	wheat	
M2	M. pasture	
M3	wheat and lupins	salt
M4	wheat	

Table 6.6   The class descriptions of the six selected training areas.

1	lupins H2
2	pasture H10
3	wheat H7
4	saltbush
5	remnant vegetation
6	salt damage

The six training areas

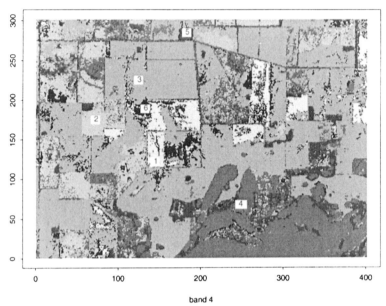

band 4

Figure 6.4   The six training areas. 1 – lupins, 2 – pasture, 3 – wheat, 4 – saltbush, 5 – remnant vegetation, 6 – salt damage. North is toward the left of the page and the ground–truth map covers a region starting from the road intersection in the northeast of the image (at approximately (5,275)).

Table 6.7   A numbering of the discriminant functions (that is the units for a full task–based MLP) together with (column 2) the classes which the discriminant function is designed to separate.

unit number	was designed to separate classes
1	1,2
2	1,3
3	1,4
⋮	⋮
15	5,6

6.4.1 Fitting a task–based MLP

A full task–based MLP was fitted to this training data set. As there are six classes, there are $\binom{6}{2} = 15$ tasks and this results in 15 pairwise–separating hyperplanes. This gives a network of size 6.15.6, with the Ω matrix of size 15×7. We will refer to this fitted model as model `mlp.6.15.6` and number the 15 tasks in the manner indicated in Table 6.7.

Early stopping was investigated using a test and validation set (see note 8, p. 65) for model `mlp.6.15.6`, with each training region divided into two regions[3]. The results indicated that, as is usually the case, the training set error rate continues to decline while the test set error rate reaches a minimum and then increases. However, in this case the effect is small (Figure 6.5) and it was decided to simply train to completion[4] on the full training set.

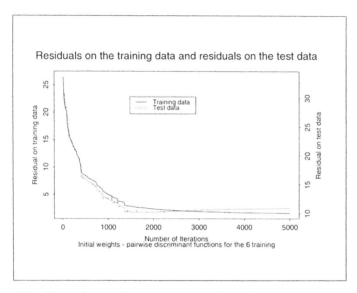

Figure 6.5   The residuals on the training data and the residuals on the test data plotted against the number of iterations. There is little evidence of over–training.

6.4.2 Pruning a task–based MLP

We assume that the training data are ordered by class. The first step in the pruning process is to take the hidden–layer outputs as a matrix, of size $N \times H$, and form the T matrix, of size $Q \times H$, that gives the number of outputs ≥ 0.5 for each class.

---

[3]Due to the spreading of light from a point source, the spectral values of neighboring pixels are correlated. As this will be true of the whole image, it makes sense to preserve this correlation structure in the training and test sets, which are therefore chosen as connected regions, rather than randomly.

[4]Training was continued until either the maximum allowed iterations or the tolerance limit was reached.

An inspection of the T matrix[5] for mlp.6.15.6 shows that each unit is performing a number of other tasks in addition to its designated task and the pruning algorithm selects the units $\{2, 8, 10, 13, 15\}$ as the indicated pruning. However, as the algorithm needs an arbitrarily set tolerance level, and breaks ties arbitrarily, it may be the case that our selection from the set of possible prunings is not optimal. It may also be the case that there are different prunings that may achieve the same outcome. As we wish to see what effects different pruning schedules will have on the final model, we decided to select a number of alternate prunings, by choosing different tie breakings, and look at several of these in some detail. The relationships between the four selected models and the units included in each model are described in Figure 6.6. The units are always numbered in the same way as in Table 6.7 (p. 80). We discuss the four models further in the next section.

6.4.3   Discussion

We consider the T matrices for these 4 models before training (the T matrices are not given but may be constructed by selecting the appropriate columns of the mlp.6.15.6$T matrix, which is given in Table 6.8). Note that while the two matrices, mlp.6.6.6.n1$T and mlp.6.6.6.n2$T, differ in many places, if they are considered in terms of their structure, that is, which groups are on which side of the hyperplanes, then they are the same except for column 6. However, column 6 in matrix mlp.6.6.6.n1$T is simply the obverse of column 6 of mlp.6.6.6.n2$T in that it has the identical structure except with values of 121 instead of 0 and vice versa[6]. Hence mlp.6.6.6.n1 and mlp.6.6.6.n2 are basically the same model up to a trivial mapping. mlp.6.5.6.n3 and mlp.6.5.6.n4 have more substantial differences, due to the fact that units 7 (in mlp.6.5.6.n3) and 15 (in mlp.6.5.6.n4) are significantly different, as indicated in the T matrix for mlp.6.15.6.

Table 6.8   The T matrix for the model mlp.6.15.6

class	hidden–layer unit

class	1	2	3	4	5	6	7	8	9	10	11	12	13	14	15
1	121	121	120	121	119	0	0	12	0	121	0	30	0	0	0
2	0	0	0	0	0	121	120	121	121	0	0	121	117	121	121
3	25	0	2	121	28	0	0	1	120	121	0	120	0	121	1
4	0	0	2	0	0	2	0	116	120	0	0	51	120	121	1
5	0	0	0	0	0	104	9	2	121	0	0	118	2	121	114
6	121	121	121	121	2	0	0	117	1	4	0	1	89	0	0

After training, the standard pruning technique was applied to each of the four models. The four T matrices for the prunings are given in Tables 6.9 to 6.12. It turns out that mlp.6.5.6.n2 and mlp.6.5.6.n4 cannot be pruned further without increasing the tolerance. mlp.6.6.6.n1 and mlp.6.5.6.n3 were pruned to

[5]The T matrix is referred to as mlp.6.15.6$T in a notation derived from the S language.
[6]This is not surprising as units 4 and 14 are also the obverse of each other.

Figure 6.6   The hierarchy of reduced models starting from mlp.6.15.6

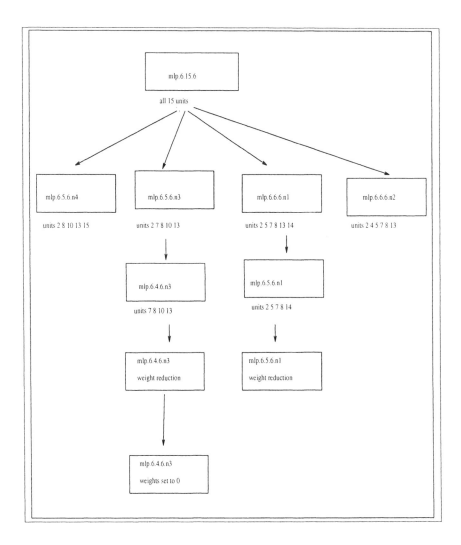

Table 6.9   The T matrix for model mlp.6.6.6.n1

class	hidden–layer unit					
	1	2	3	4	5	6
1	0	117	0	0	121	0
2	120	0	121	109	0	120
3	0	34	0	0	0	0
4	120	0	0	120	0	121
5	0	0	5	0	0	121
6	114	1	0	78	121	0

Table 6.10   The T matrix for model mlp.6.5.6.n4

class	hidden–layer unit				
	1	2	3	4	5
1	121	0	121	0	0
2	0	121	0	120	121
3	0	0	121	0	0
4	0	14	0	120	2
5	0	0	0	4	109
6	120	114	4	52	0

Table 6.11   The T matrix for model mlp.6.6.6.n2

class	hidden–layer unit					
	1	2	3	4	5	6
1	0	121	0	0	121	121
2	120	0	121	106	0	0
3	0	75	0	0	0	121
4	120	0	0	120	0	0
5	0	0	2	0	0	0
6	114	5	0	74	121	121

Table 6.12   The T matrix for model mlp.6.5.6.n3

class	hidden–layer unit				
	1	2	3	4	5
1	0	0	0	121	121
2	120	121	0	0	0
3	0	0	0	0	121
4	121	0	112	0	0
5	1	10	0	0	0
6	116	0	0	120	4

Table 6.13   The action of the hidden units of the mlp.6.5.6.n1 model in terms of which classes are on the "+" side of the separating hyperplanes. "Some wheat" means that the separating hyperplane bisects the wheat training class.

unit number	gives a "+" value for classes
hidden unit 1	pasture, salt bush and salt damage
hidden unit 2	lupins and some wheat
hidden unit 3	pasture
hidden unit 4	lupins and salt damage
hidden unit 5	pasture, salt bush and remnant vegetation

mlp.6.5.6.n1 and mlp.6.4.6.n3 respectively and further training was conducted on these models.

6.4.4   Analysis of the Υ matrices

After training, we consider the two pruned models, mlp.6.4.6.n3 and mlp.6.5.6.n1, to see if any further pruning or simplification is possible.

The outputs of the hidden–layer units may be analyzed via one of two means. For the training data, we can take the raw outputs or the T matrix simplification. Also, for the whole image, the hidden–layer outputs can themselves be interpreted as $[0, 1]$ greyscale images. Using the T matrix as a guide, we can determine what the hidden–layer units are doing in terms of class separations. For two of the models (mlp.6.4.6.n3 and mlp.6.5.6.n1), Tables 6.13 and 6.14 interpret the T matrices (given, respectively, in Tables 6.15 and 6.16) to show which classes are on the "+" side of the separating hyperplane.

Table 6.14   The action of the hidden units of the mlp.6.4.6.n3 model in terms of which classes are on the "+" side of the separating hyperplanes.

unit number	gives a "+" value for classes
hidden unit 1	pasture, salt bush and salt damage
hidden unit 2	pasture
hidden unit 3	lupins and salt damage
hidden unit 4	lupins and wheat

As the next stage in the MLP model, the rows of the Υ matrix indicate the linear combinations of the hidden–layer outputs (and thus of the columns of the T matrices) that are important in determining the various class labels. However, an inspection of the Υ matrix, for either model, indicates a complex pattern involving all of the hidden–layer units.

6.4.5   Weight simplification

In order to improve the interpretability of the Υ matrix, we have employed a weight elimination penalty term (see note 6, p. 64). This implementation, unlike the

Table 6.15   The T matrix for model mlp.6.4.6.n3

class	1	2	3	4
1	0	0	121	121
2	120	121	0	0
3	0	0	0	121
4	120	0	0	0
5	0	6	0	0
6	116	0	120	5

Table 6.16   The T matrix for model mlp.6.5.6.n1

	hidden–layer unit				
class	1	2	3	4	5
1	0	118	0	0	0
2	120	0	121	115	120
3	0	34	0	0	0
4	120	0	0	120	121
5	0	0	5	0	121
6	114	. 1	0	12	0

standard one, is only applied to the Υ matrix, not to both matrices. Ideally, λ should be chosen by cross–validation; however, we have simply set it to 0.01 on the basis of previous experience. Ripley (1996, §5.5) investigates this question and suggests choosing λ in the range $0.001 - 0.1$ and comments that the value for λ is not critical within a factor of 5.

After weight elimination was carried out on the mlp.6.4.6.n3 model, the weights were simplified so that weights of magnitude less than 0.5 were fixed at 0, weights greater than 3 were set to 4, and weights less than -3 were set to -4. This had no effect on the accuracy of the fit to the training data (see the classification matrix in Table 6.17) and gave the following Ω matrix:

$$\begin{bmatrix} -2.23 & -4 & 0 & 4 & 0 \\ -2.33 & 0 & 4 & 0 & 0 \\ -1.78 & 0 & 0 & -4 & 4 \\ -1.45 & 4 & -4 & -4 & 0 \\ 1.56 & -4 & 0 & 0 & -4 \\ -1.72 & 0 & 0 & 4 & -4 \end{bmatrix}.$$

We selected two output units for analysis[7]. Output unit 5 (which detects remnant vegetation) is using minus the sum of hidden–layer units 1 and 4. An inspection of Table 6.14 reveals that these two units between them have all the classes except remnant vegetation on the "+" side of their hyperplanes. Hence remnant vegetation is the only class left on the "−" side; minus the sum of hidden units 1 and 5 thus isolates remnant vegetation on the "+" side of output unit 5.

---

[7]The hidden–layer bias unit is numbered as unit 0.

Table 6.17   The classification matrix for the model mlp.6.4.6.n3.reduced. The true
class is given down the table and the assigned class across the table.

class	1	2	3	4	5	6
1	121	0	0	0	0	0
2	0	121	0	0	0	0
3	0	0	121	0	0	0
4	0	0	0	121	0	0
5	0	0	0	3	118	0
6	4	0	1	0	0	116

Output unit 1 (which identifies lupins) is formed by differencing hidden–layer
units 1 and 3, which respectively separate pasture, salt–bush and salt damage
(hidden unit 1) and lupins and salt damage (hidden unit 3). Hence unit 1 is
receiving three sets of input signals: positive signals (for lupins); negative signals
(for pasture and salt–bush); and signals near 0 for salt damage.

6.5   EVALUATING THE MODELS

The methods discussed in Chapter 5 (p. 53) are available to evaluate the model fits.
However, it may be the case that the differences between models, when applied to
the whole image, are unacceptably great for reasons other than that expressed in
the class-conditional error rates.

To see if this is the case, we assume that the mlp.6.15.6 model gives a satisfactory classification map and track the number of pixels which change labels for a
series of consecutively pruned models. The results are given in Table 6.18, which
shows that the greatest difference is of the order of 7%.

Table 6.18   The percentage of pixels changing from one model to the next.

	mlp.6.5.6.n3	mlp.6.4.6.n3	mlp.6.4.6.n3. reduced	mlp.6.4.6.n3. reduced.2
mlp.6.5.6.n3		3.56	5.96	6.75
mlp.6.4.6.n3			3.50	4.19
mlp.6.4.6.n3.reduced				2.66

However, there is a significant difference between a random selection of 7% of the
pixels changing, and a single 7% block of the image (e.g., a field) changing class.
The first will make little difference to the interpretation of the image (although
the image may look degraded), whereas the second is likely to be of more serious
concern.

We can represent the changes between models as black and white images showing
the pixels that have changed labels. This serves as a warning device. As long as
the pruning sequence is only changing a random pattern of pixel classifications, it
may be safe to proceed. However, when a connected region, such as a field, changes

its classification, it is advisable to stop the pruning process and examine both the model, and the data that are changing classification, more closely.

The difference image for the pruning `mlp.6.15.6` to `mlp.6.6.6.n1` is given in Figure 6.7. While some small and irregularly shaped regions have changed classification, the majority of the changes have been within the classes of saltbush and remnant vegetation. A sequence of such images indicates that the given prunings have not altered the classification of any of the crop or pasture areas or of the salt damaged areas (which were the main focus of interest in the study). This example is taken up again in Section 7.6, p. 109.

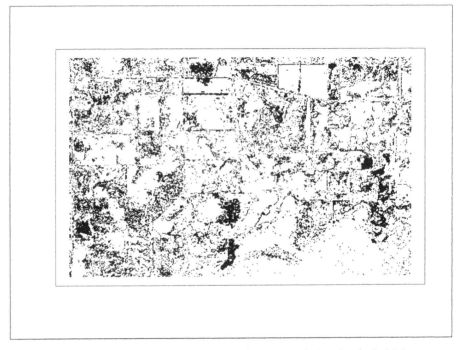

Figure 6.7   The differences between the classified images mlp.6.15.6 and mlp.6.5.6.

6.6 CONCLUSION

This analysis has shown that task-based MLPs and task-based pruning can produce effective MLPs of reasonable size. Figure 6.8 gives the final classified image produced by the model `mlp.6.5.6.n4`.

However, we have also seen that there are a number of equally acceptable prunings giving rise to a number of different final models. In addition, we have seen that other members of an acceptable set of prunings may give rise to more parsimonious models than the one selected by the task–based pruning algorithm. Hence it may be advisable to search the space of possible prunings rather than simply selecting one. In the past, we have avoided this due to the heavy computational burden that it entails.

We have also seen that the outputs of the hidden–layer units can be interpreted and displayed as images, which leads to a greater understanding of the actions of the MLP. The Υ matrix defines linear combinations of the hidden–layer images and these combinations can be interpreted as adding and differencing the hidden–layer images to identify the classes of interest. It appears that it may be feasible to reduce the Υ matrix to simpler integer combinations of the hidden–layer units and still retain the same classification accuracy.

The fact that the activation functions are continuous accounts for the large amount of information present in the hidden–layer and output-layer images. Although we have not used the fact in this chapter, it is possible to interpret the output of both hidden- and output-layers as images. See Figure 7.2 (p. 102) where the Martin's farm example is considered further.

We have found that, despite the fact that in the task–based MLP each hidden–layer unit is designed to separate two super–classes, a very tight pruning (such as mlp.6.4.6) may result in a hidden–layer unit acting like a ternary logic unit, in that it may separate three super-classes, rather than a binary logic unit. This may be considered undesirable as, while the final model is simple in terms of the number of units, its action may be more complex.

The mlp S library contains functions that facilitate fitting and pruning task-based networks. See Dunne et al. (1992) for an application of task–based pruning to a speaker-independent vowel recognition problem.

6.7 COMPLEMENTS AND EXERCISES

6.1 We can select the initial estimates of the Υ matrix in order to decrease the training time. If X (of size $N \times P + 1$) is the data matrix of x values, then the matrix Υ is chosen so that

$$\underset{Q \times (H+1)}{\Upsilon} [1, \{f_h(\underset{(P+1) \times N}{\Omega \ X^T})\}^T]^T \approx \underset{Q \times N}{B^T},$$

where B is a matrix formed from the target matrix T such that

$$B = \begin{cases} 1 & \text{where } T = 1, \text{and} \\ -1 & \text{where } T = 0. \end{cases} \tag{6.1}$$

In general, if we write $M = [1, \{f_h(\Omega X^T)\}^T]$ then taking the singular-value decomposition of M,

$$\underset{N \times (H+1)}{M} = \underset{N \times (H+1)}{U} \underset{(H+1) \times (H+1)}{\text{diag}(\lambda)} \underset{(H+1) \times (H+1)}{V^T}$$

we can form a generalized inverse of M,

$$\underset{(H+1) \times N}{M^-} = \underset{(H+1) \times (H+1)}{V} \underset{(H+1) \times (H+1)}{\text{diag}(\lambda)^{-1}} \underset{(H+1) \times N}{U^T}.$$

If any of the λ are too small then the stability of the solution can be improved by setting small singular values to 0 (Strang, 1988). Then

$$\underset{q \times (H+1)}{\Upsilon} = \underset{q \times N}{B^T} \underset{N \times (H+1)}{M^-}.$$

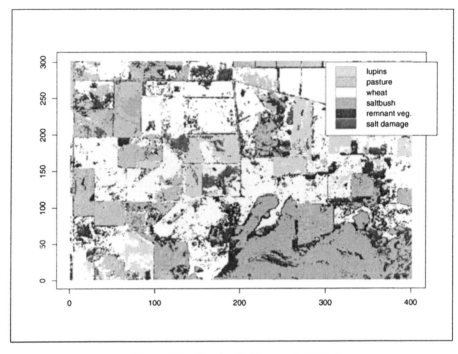

Figure 6.8   The classified image mlp.6.5.6.n4.

In the case where the MLP is used for function approximation, rather than classification, the discussion given here could be used as the basis for a hybrid training strategy where Υ is estimated via a linear equation at each step. This has been considered in Webb et al. (1988) and Barton (1991).

We note that (6.1) means that $\{f_Q(\Upsilon M^\tau)\}^\tau$ will not be a good approximation to T and that the approximation could be improved by setting $B \leftarrow mB$ where $m > 1$. However, in general it is desirable to restrain, rather than encourage, the growth of the weights (see Section 5.4 (p. 62) for a discussion and survey of the literature on penalized training).

6.2   The Hilbert matrix has i, j entries equal to $1/(i + j - 1)$. We consider H_6 which is considered to be "nearly singular" (Noble and Daniel, 1988).

Calculate the condition number (p. 62). Investigate the effect on the inverse of a slight perturbation on one or more entries in the matrix.

Calculate the distance of the 6th column of H_6 from the subspace spanned by the fist 5 columns.

6.3   The mlp library provides the following function:

pd calculates the Ω weights for the MLP model as the discriminant functions for each pair of classes;

calcu calculates the appropriate Υ matrix;

`mlp.diagnostic` calculates a number of diagnostic measure (as discussed in this chapter);

`prune` prunes the weights based on these diagnostic measures.

Set up a synthetic data set of 4 Gaussians in $\mathbb{R}^2$ with means at $(\pm 3, \pm 3)$ and calculate the pairwise discriminant functions as starting values. Prune the network down to the smallest applicable and plot the separating hyperplanes.

6.4 Do the same with the Iris data set from Problem 2.2, p. 18.

6.5 The Martin's farm data set is available in the `mlp` library.

```
data("martins_farm_august_1990")
dim(maraug90)
#[1] 120701      6
```

The columns of `maraug90` are the first six bands of the table 6.4. We can extract one of the bands and reshape it to display it as an image.

```
temp<-maraug90[,4]
dim(temp)<-c(301,401)
temp<-t(temp)
image(temp)
```

The 6 training areas can be selected as follows:

```
select<-function(data,a,b,c,d){
temp<-matrix(0,(b-a+1)*(d-c+1),6)
tt<-1
for( i in a:b){
for (j in c:d){
temp[tt,]<-data[(i-1)*301 +j,]
tt<-tt+1
temp
}}}
```

```
train1<-select(image,140,150,117,127)   #train1   H2 lupins
train2<-select(image,68,78,170,180)      #train2   H10 pasture
train3<-select(image,120,130,220,230)    #train3   H7 wheat
train4<-select(image,242,253,63,73)      #train4   saltbush
train5<-select(image,178,188,280,290)    #train5   remnant vegetation
train6<-select(image,128,138,184,194)    #train6   salt damage
```

Just as a check, we can plot the training areas on the image and see that they are reasonable.

```
temp[140:150,117:127]<-0
temp[68:78,170:180]<-0
temp[120:130,220:230]<-0
temp[242:256,63:73]<-0
temp[178:188,280:290]<-0
temp[128:138,184:194]<-0
```

```
image(1:401,1:301,temp)
title(main="The 6 training areas",sub="band 4")
text(146,123,"1")
text(74,176,"2")
text(126,226,"3")
text(249,69,"4")
text(184,286,"5")
text(134,190,"6")

par(pin=c(4,3))
legend.names<-c("H2 lupins","H10 pasture","H7 wheat","saltbush",
 "remnant vegetation","salt damage")
legend(295.1943,246.3558,legend.names,pch="123456")
```

Repeat the analysis of this chapter with a cross–entropy penalty function and a softmax activation function at the output layer. You will find that while many of the details differed for the two analyses, the same general pattern emerges.

6.6   Using the lda function from the MASS library (or otherwise), calculate the LDA (Chapter 3, p. 19) for the training data and produce a classified image.

Using Problem 3.9 (p. 34) calculate the typicality probability for each pixel and make a typicality image. It may be the case that some regions of the image are found to have a low typicality. Consider the fact that the 6 classes may not in fact capture all of the variability in the image.

It is a common experience with this sort of data to have to add unknown classes during the course of the analysis.

CHAPTER 7

INCORPORATING SPATIAL INFORMATION INTO AN MLP CLASSIFIER

7.1 ALLOCATION AND NEIGHBOR INFORMATION

We continue with the theme of the major example from the previous chapter, that is, the analysis of multispectral image data. The classification of such data where the pixel labels are spatially related may be improved by modeling the local dependence. We consider extending the Multi–layer Perceptron model to incorporate this spatial information.

A Markov Random Field (MRF) is a model in which a label is assumed to be conditional on the labels of the neighboring pixels only. The posterior probability of a class label given the spectral values and neighboring labels can then be maximized by an iterative process depending only on local information. It is shown that an iterative updating procedure can be implemented for MLP models. This is described in terms of a MRF model and it is shown that this is related to a Hopfield network. Two examples are considered, one using synthetic data and the other a remotely sensed image of an agricultural property (the Martin's farm example from Chapter 6, p. 69).

The event of belonging to a class q is designated by $g = g^q$ or just g^q for $q = 1, \ldots, Q$ classes. In a classification or discrimination problem we allocate x to the class for which $P(g^q|x)$, the probability of the class given the observation, is a maximum. If $P(g^q)$ denotes the prior probability of membership of class g^q and $P(x|g^q)$ the density for class g^q, then labels can be chosen to maximize $P(g^q|x)$ via

Bayes' formula

$$P(g^q|x) = \frac{P(g^q)P(x|g^q)}{\sum_{q'} P(g^{q'})P(x|g^{q'})}$$
$$\propto \quad P(g^q)P(x|g^q). \tag{7.1}$$

Hence the classification problem is resolved into estimating $P(g^q)$ and $P(x|g^q)$ (this is within the sampling paradigm as discussed in Section 2.4, p. 15).

Prior information may determine the choice of $P(g^q)$ or, in the absence of such information, it may be set to $1/Q$ for all of the classes. For $P(x|g^q)$, distributional assumptions are usually made and the parameters of the distribution are estimated from a sample; for example, if a Gaussian distribution is assumed for $P(x|g^q)$, then

$$P(x|g^q) \propto \quad \exp\{-\tfrac{1}{2}(x-\mu_q)^T \Sigma_q^{-1}(x-\mu_q)\}$$
$$= \quad e^{-\frac{1}{2}D_q^2(x)},$$

where $D_q^2(x)$ is the squared Mahalanobis distance of x from the mean of class g^q. In order to apply equation (7.1) it is generally necessary to estimate μ and Σ from a set of training data.

As it was shown in Section 4.2 (p. 35) that the standard MLP directly estimates the posterior probabilities $P(g^q|x)$ directly, it might seem that equation (7.1) is not relevant when we are using an MLP model. However, (7.1) has implications for extending the MLP model to allow for spatial dependence.

7.1.1   Sources and extent of spatial dependence

For remotely sensed images, a P–dimensional vector of spectral measurements is available for every pixel in the image. The nature of the remote sensing process is such that the values recorded at a given pixel are affected by the reflectances of the neighboring pixels (via the "point spread function"). Hence, for groups of neighboring pixels, the spectral values are typically correlated.

Campbell and Kiiveri (1988) considered the magnitude of the spatial correlation of the spectral values within homogeneous spectral classes for Landsat MSS and TM data and for AVHRR data[1]. They found that the correlations are typically of the order of 0.4 to 0.6 for neighboring pixels with the diagonally adjacent pixels showing less correlation than the horizontally and vertically adjacent pixels. There is generally an exponential decrease with distance. However, the usual procedure, both for maximum likelihood classification and for MLP models, is to ignore any correlation between data points in the training data.

In addition to the correlation of the spectral values, it is typically the case with remotely sensed image data that there are many spatially contiguous pixels belonging to the same class. Hence there may be some contextual information in the labels of neighboring pixels which could be used to improve confidence in the

---

[1]The Landsat series of satellites has carried two sensors, the 4-band Multi-Spectral Scanner (MSS) instrument with a pixel size of 100 meters and the 7-band Thematic Mapper (TM) instrument with a pixel size of 25 meters. The NOAA satellite carries the Advanced Very High Resolution Radiometer (AVHRR) instrument, which has a pixel size of 1 kilometer but has high spectral resolution.

ascribed pixel label. It is this spatial information that we are concerned with in this chapter.

7.1.2 Extending the model

To extend model (7.1) to take account of the information in neighboring pixels, it is necessary to impose more structure on $P(g^q)$ and on $P(x|g^q)$. Following the notation of Kiiveri and Campbell (1992), the spectral variables are modeled as an $m \times n$ array of $P \times 1$ random vectors X_i measured at the sites of a rectangular lattice $\mathbf{I} = \{(r, s) : 1 \leq r \leq m, 1 \leq s \leq n\}$. Associated with each pixel is a class label G_i so that $\mathbf{X} = \{X_i : i \in \mathbf{I}\}$ and $\mathbf{G} = \{G_i : i \in \mathbf{I}\}$. Lowercase letters denote realizations of random variables.

The notation is modified slightly in this chapter as we now need to indicate the spatial position of an observation as well as its number and class. Each pixel in the interior of the image has four "first–order" neighbors (the horizontally and vertically adjacent pixels), and eight "second–order" neighbors. More generally, if some set of cliques is defined as constituting the neighborhood of each pixel, then the neighbors can be indexed by $n(i) = \{j = 1, \ldots, N(i) | j$ is a neighbor of $i\}$ (with the central pixel numbered 0) as illustrated in Figure 7.1. For simplicity, for most of this discussion "first–order" neighborhoods are considered. We let $\bar{n}(i)$ denote the pixel i together with its neighbors, and in addition, let $r(i)$ denote the set $\mathbf{I}\backslash\{i\}$ (the rest of the pixels).

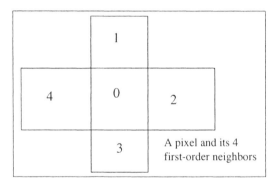

Figure 7.1   Each pixel in the interior of the image has four "first–order" neighbors (the horizontally and vertically adjacent pixels). The neighbors can be indexed by $n(i) = \{j = 1, \ldots, N(i) | j$ is a neighbor of $i\}$, with the central pixel numbered 0.

Two assumptions are made,

$$P(g_i|g_{r(i)}) = P(g_i|g_{n(i)}) \qquad (7.2)$$

and

$$P(x_i|x_{r(i)}, \mathbf{g}) = P(x_i|g_i). \qquad (7.3)$$

Assumption (7.2) means that each label depends on the other labels only through the labels of the neighboring pixels; thus the labels are *a priori* a realization from a Markov random field. In Section 7.2, p. 98, we give a brief discussion of Markov random fields, and for further details see Geman and Geman (1984), Besag (1986) and Kiiveri and Campbell (1992).

Under assumption (7.3), the spectral values are conditionally independent given the labels so the effect of the point spread function is ignored. Kiiveri and Campbell (1992) develop an argument using only the weaker assumption that $P(x_i|x_{r(i)}, \mathbf{g}) = P(x_i|x_{n(i)}, g_{\bar{n}(i)})$, so that for a given $\mathbf{g}$, the spectral data are also a realization from a Markov random field. However, they found that the improvement in classification accuracy was small compared with the effect of incorporating neighbor context via assumption (7.2).

7.1.2.1 *Modeling the spectral values*

For MLP models, the correlation of the spectral values of neighboring pixels may not be critically important, as the model makes no distributional assumptions about the input variables. However, for Maximum Likelihood estimation, it is important to adequately model the distribution $P(x|g)$. For example, if it is is modeled as a Gaussian distribution, then the correlations of the spectral variables, due to contiguous pixels, conflict with one of the modeling assumptions and we get a biased (inflated) estimate of Σ. Campbell and Kiiveri (1993) have proposed changes to the usual discriminant analysis procedure to take into account the spatial correlation of the data.

We consider the question of modeling the spectral correlation principally in order to generate simulated data (scenes) with realistic properties to test proposed methods of classification, and also to understand the process by which data sets such as remotely sensed images are generated. Let the mean and covariance of X_i be μ_i and Σ_i for $i \in \mathbf{I}$. Following Kiiveri and Campbell (1992), we describe the standardized variables

$$Y_i = \Sigma_i^{-\frac{1}{2}}(X_i - \mu_i)$$

by a (multivariate) conditional autoregressive (CAR) model. Under this model,

$$E[Y_i|Y_{r(i)}] = \sum_{j \in n(i)} b_{i-j} Y_j,$$

that is, the expected value of Y_i is a linear combination of the values of its neighbors $(b_{i-j} = b_{j-i})$, and

$$V[Y_i|Y_{r(i)}] = b_0^{-1} I$$

for some set of parameters $\{b_0, b_{i-j}$ for $j \in n(i)\}$. The model leads to a block structure on $\Sigma^{-1}(g)$, the covariance for $\mathbf{X}$ such that, for fixed g and each i, X_i is conditionally independent of the rest of the variables given its neighbors $X_{n(i)}$. See Kiiveri and Campbell (1992) for further details.

The example in Problem 7.1 (p. 114) provides such a simulated scene. See also Wilson (1992) for a discussion of the creation of auto–correlated multispectral maps. Note that Wilson's method produces the same level of correlation for each spectral band and every group cover type, something the method described in Kiiveri and Campbell avoids.

7.1.2.2 *Modeling the prior distribution of the labels*

$P(g_i|g_r)$ may be modeled as

$$P(g_i|g_r) = P(g_i|g_{n(i)})$$
$$= c(g_{n(i)}) \exp\{\alpha_i + \beta_h N_h(g,i) + \beta_v N_v(g,i) + \beta_d N_d(g,i)\},$$

where $c(g_{n(i)})$ is a normalizing constant, $N_h(g,i)$ denotes the number of horizontal neighbors of site i with the same label as g_i, and N_v and N_d perform corresponding tasks for the vertical and diagonal neighbors. This model allows for the spatial dependence to be different in different directions. Alternatively, if we do not expect any directional effects, it might be modeled as

$$P(g_i|g_{r(i)}) = c(g_{n(i)}) \exp\{\alpha_i + \beta N(g,i)\} \tag{7.4}$$

where $N(g,i)$ is the total number of neighboring pixels with the same label as g.

7.1.3 A Markov model for the allocation of pixels

Using assumptions (7.2) and (7.3), it can be shown that the posterior distribution inherits the conditional independence constraints so that

$$P(g_i|\mathbf{x}, g_{r(i)}) = P(g_i|x_{\bar{n}(i)}, g_{n(i)}).$$

For all $i \in \mathbf{I}$, suppose that $P(x_{r(i)}|\mathbf{g})$ does not depend on g_i. Then

$$
\begin{aligned}
P(g_i|\mathbf{x}, g_{r(i)}) &= \frac{P(\mathbf{x}, g_{r(i)}, g_i)}{\sum_{g_i} P(\mathbf{x}, g_{r(i)}, g_i)} \\
&= \frac{P(x_i|x_{r(i)}, \mathbf{g})P(x_{r(i)}|\mathbf{g})P(g_i|g_r)P(g_r)}{\sum_{g_i} P(x_i|x_{r(i)}, \mathbf{g})P(x_{r(i)}|\mathbf{g})P(g_i|g_r)P(g_r)} \\
&= \frac{P(x_i|x_{n(i)}, g_{\bar{n}(i)})P(x_{r(i)}|\mathbf{g})P(g_i|g_{n(i)})P(g_r)}{\sum_{g_i} P(x_i|x_{n(i)}, g_{\bar{n}(i)})P(x_{r(i)}|\mathbf{g})P(g_i|g_{n(i)})P(g_r)}
\end{aligned}
$$

and, using the above supposition,

$$
\begin{aligned}
P(g_i|\mathbf{x}, g_{r(i)}) &= \frac{P(x_i|x_{n(i)}, g_{\bar{n}(i)})P(g_i|g_{n(i)})}{\sum_{g_i} P(x_i|x_{n(i)}g_{\bar{n}(i)})P(g_i|g_{n(i)})} \\
&= P(g_i|x_{\bar{n}(i)}, g_{n(i)}).
\end{aligned}
$$

This forms the basis for the pixel allocation procedures, developed in Besag (1986) and Geman and Geman (1984), based on the idea of constructing a sequence $\mathbf{g}^{(m)}$ of labels such that $P(\mathbf{g}^{(m)}|\mathbf{x})$ is increasing. When pixel i is visited, $P(g_i|x_{\bar{n}(i)}, g_{n(i)})$ is maximized by choosing the next label in the sequence $g_i^{(m)}$ such that, for $g_{n(i)}$ fixed,

$$P(g_i^{\text{new}}|x_{\bar{n}(i)}, g_{n(i)}) \geq P(g_i^{\text{old}}|x_{\bar{n}(i)}, g_{n(i)}), \tag{7.5}$$

where g^{new} is the new label and g^{old} is the previous label in the sequence. Hence the algorithm finds a maximum of the *posterior* distribution.

Besag (1986) finds a local maximum by "iterative conditional modes" so that the maximum of the distribution is chosen at each iteration. Geman and Geman (1984) consider a simulated annealing algorithm for constructing the sequence $g^{(m)}$ so that, providing a "cooling schedule" is constructed appropriately, the global maximum may be found.

Kiiveri and Campbell (1992) discuss various aspects of fitting these models and compare them on a simulated data set. They conclude that the fitting procedure is very sensitive to the starting values, hence good initial estimates are very important. They also conclude that the global maximum may not be a satisfactory map. This is discussed further at the end of Section 7.2.

7.2 MARKOV RANDOM FIELDS

We have already set up the framework that we require in equations (7.2), (7.3), (7.4) and (7.5). However, as these assumptions are derived within the framework of Markov random fields (MRFs), we briefly review the area.

The assumption that $P(g_i|g_{r(i)}) = P(g_i|g_{n(i)})$ is similar to that of a Markov chain and defines a Markov random field (MRF), an extension of a Markov chain to two dimensions. The MRF can be specified in terms of the local conditional probabilities, as in equation (7.2), or in terms of the joint probability distribution of the labels. It is known that if the conditional distribution of the labels adheres to the Markov assumption (7.2) then the joint distribution is restricted by some non-obvious consistency conditions. These are described by the factorization theorem given below, following the exposition of Besag (1974). We require the further assumption that if $P(g_i) \neq 0$ for each $i \in \mathbf{I}$ then

$$P(g_1, g_2, \ldots, g_n) \neq 0, \tag{7.6}$$

known as the positivity condition. In addition we write $|\mathbf{I}| = n$.

The factorization theorem[2] asserts the following: suppose that $\mathbf{g}$ and $\mathbf{h}$ are are two possible realizations of $\mathbf{G}$ then

$$\frac{P(\mathbf{g})}{P(\mathbf{h})} = \prod_{t=1}^{n} \frac{P(g_i \mid g_1, g_2, \ldots, g_{i-1}, h_{i+1}, \ldots, h_n)}{P(h_i \mid g_1, g_2, \ldots, g_{i-1}, h_{i+1}, \ldots, h_n)}. \tag{7.7}$$

The proof is as follows. We can write

$$P(\mathbf{g}) = P(g_n \mid g_1, \ldots, g_{n-1})P(g_1, \ldots, g_{n-1}),$$

then we can introduce h_n and write

$$P(\mathbf{g}) = \frac{P(g_n \mid g_1, \ldots, g_{n-1})}{P(h_n \mid g_1, \ldots, g_{n-1})} P(g_1, \ldots, g_{n-1}, h_n)$$

and operate on g_{n-1} in the $P(g_1, \ldots, g_{n-1}, h_n)$ term, introducing h_{n-1}, to get

$$P(g_1, \ldots, g_{n-1}, h_n) = \frac{P(g_{n-1} \mid g_1, \ldots, g_{n-2}, h_n)}{P(h_{n-1} \mid g_1, \ldots, g_{n-2}, h_n)} P(g_1, \ldots, g_{n-2}, h_{n-1}, h_n).$$

Assumption (7.6) ensures that the denominator is never 0. Continuing in this way we can reduce the expression to equation (7.7). As the labeling of the sites is arbitrary, there are $n!$ factorizations of $P(\mathbf{g})/P(\mathbf{h})$ all of which must be equivalent. This means that there are severe constraints on the local conditional probability distributions $P(g|g_{n(i)})$ in order to achieve a consistent $P(\mathbf{g})$. This is discussed further at the end of the current section (p. 100).

7.2.1 The Gibbs random field

We define a Gibbs random field (GRF) as one in which the joint probability distribution is of the form

$$P(\mathbf{g}) = \frac{1}{Z} \exp\left\{ \frac{-U(\mathbf{g})}{T} \right\}, \tag{7.8}$$

---

[2]The theorem is generally attributed to Hammersly and Clifford in 1971 but was not published for twenty years after that (Ripley, 1996, §8).

known as a Gibbs distribution. Here U is an energy function (the more probable configurations are those with lower energy), T is a "temperature," and Z is a normalizing term called the "partition function." The names reflect the statistical mechanics origin of the distribution.

We assume that g takes only a finite number of values and define a *clique* as a set of pixels which either consists of a single pixel or else a set such that every pixel in the set is a neighbor of every other pixel in the set. We group the cliques according to size, so that $C_1 = \{i | i \in \mathbf{I}\}$, the cliques of size 1, and $C_2 = \{\{i, i'\} | i \in \mathbf{I}, i' \in \{n(i)\}\}$ and so on.

Hammersley and Clifford showed the most general form of U is

$$U(\mathbf{g}) = \sum_{i \in C_1} V_1(g_i) + \sum_{\{i,i'\} \in C_2} V_2(g_i, g_{i'}) + \dots \tag{7.9}$$

where the V are positive functions and symmetric in their arguments. A proof that a GRF is a MRF can be given by showing that

$$P(g_i | g_{r(i)}) = \frac{\exp(-\sum_{C \in A_i} V_C(g))}{\sum_{g'_i} \exp(-\sum_{C \in A_i} V_C(g'_i))},$$

where $g'_i \in \{g_{r(i)}\}$ and A_i is the set of all cliques containing i (Li, 1995). Hence knowing U allows the conditional probabilities to be recovered. A proof in the other direction is more involved, see Ripley (1996, §8).

A special case of the GRF is the class of auto-models where

$$U(\mathbf{g}) = \sum_{i \in C_1} g_i \beta_i + \sum_{\{i,i'\} \in C_2} g_i g_{i'} \beta_{i,i'} + \dots$$

of which we consider some specific examples. Consider a "multi-color" MRF where we have class labels $k \in \{1, \dots, Q\}$ but without an intrinsic ordering of the labels. For $Q = 3$, for example, we can set

$$P(\mathbf{g}) \propto \exp(\alpha_1 n_1 + \alpha_2 n_2 + \alpha_3 n_3 - \beta_{12} n_{12} - \beta_{13} n_{13} - \beta_{23} n_{23}),$$

where n_1 is the number of pixels with label 1, n_{12} is the number of $\{g^1, g^2\}$ neighbor pairs etc, so we are assuming that only pairs of sites are neighbors. By consideration of realizations which differ only at pixel i, it can be shown that

$$P(g_i^1) \propto \exp(\alpha_1 - \beta_{12} n_{12} - \beta_{13} n_{13})$$

or, if we let $\beta = \beta_{12} = \beta_{13}$,

$$P(g_i) \propto \exp\{\alpha_1 + \beta N(g, i)\}, \tag{7.10}$$

which is a similar model to (7.4), differing in that (7.4) allows a separate α_i for each pixel.

We can also consider a set of l ordered values such as a grey–level image with intensity values of 0 (black) to 255 (white). This gives an auto–binomial model with

$$P(g_i | g_{n(i)}) \sim B(l, \Theta),$$

where

$$\Theta = \frac{\exp(\alpha + \sum_{n(i)} \beta_{n(i)} g_{n(i)})}{1 + \exp(\alpha + \sum_{n(i)} \beta_{n(i)} g_{n(i)})}.$$

In the case where we have two classes, labeled 0 and 1, the multi–color and the ordered MRF reduce to the same model, the auto–logistic or Ising spin glass model with

$$U(x) = \sum_i \alpha_i g_i + \sum_i \sum_{n(i)} \beta_{i,n(i)} g_i g_{n(i)}$$

and

$$P(g_i^1 \mid g_{n(i)}) = \frac{\exp(\alpha_i + \sum_{n(i)} \beta_{n(i)} g_{n(i)})}{1 + \exp(\alpha_i + \sum_{n(i)} \beta_{n(i)} g_{n(i)})}. \tag{7.11}$$

It is also possible to construct models where the class label is a continuous variable, such as the auto–normal model of Besag (1986).

Geman and Geman (1984) comment that (7.9) and (7.8) allow the MRF to be specified in terms of the potential rather than in terms of the local conditional distributions, which is difficult to do in such a way that the conditions of result (7.7) are met. However, it is not clear that the MRF model is more than locally and approximately correct in image restoration problems. Besag (1986) warns that while (7.2) mimics the local characteristics of the scene, it may be that the global properties of the assumed MRF will dominate the restoration of the image, producing long–scale dependencies between pixels and a poor reconstruction. In addition, Besag questions whether replacing a local conditional distribution based on $\exp\{\beta N(g,i)\}$ with one based on $\beta N(g,i)$ will lead to a poorer reconstruction, although it will lead to a violation of the consistency of the MRF.

7.3 HOPFIELD NETWORKS

The Hopfield network is closely associated with the insights of the Hebbian learning rule introduced by Hebb (1949). This is the idea that "when . . . cell A is near enough to excite cell B and repeatedly or persistently takes part in firing it, some growth process or metabolic process takes part in one or both cells such that A's efficiency as one of the cells firing B is increased" (Hebb quoted in Serra and Zanarini, 1990).

A Hopfield network consists of N interconnected units, each of which is a binary logic device with the two states 0 and 1, say. The state of the ith unit is X_i. The connections between the units, which we describe by a set of weights $\{w\}$, are real valued and are modified by a learning rule. Each unit receives input from the other units so that the input to the ith unit at a fixed time is $\Phi(X_i) = \sum_{j=1}^{N} w_{ij} X_j$, the weighted sum of the states of the set of units. This input is compared with a threshold θ and the following action is taken:

$$
\begin{array}{lll}
X_i(t+1) = 1 & \text{if} & \Phi\{X_i(t)\} > \theta \\
X_i(t+1) = 0 & \text{if} & \Phi\{X_i(t)\} < \theta \\
X_i(t+1) = X_i(t) & \text{if} & \Phi\{X_i(t)\} = \theta
\end{array}
$$

If we introduce a probability distribution on $\Phi\{X_i(t)\}$, as in the Boltzman machine, then we have a close relationship with the MRF models. For a Boltzman machine, X_i is a Bernoulli random variable with a logistic distribution (equation 2.2, p. 11)

$$P(X_i = 1) = \text{logistic}[\Phi\{X_i(t)\}].$$

If we define neighborhood cliques so that $w_{ij} = 0$ if i and j are not neighbors, this model corresponds to the MRF model (7.11). The Hopfield network has been widely explored for optimization problems, such as the traveling salesperson. For further details on it and the Boltzmann machine see Pao (1989), Serra and Zanarini (1990) or Lippmann (1987) for a brief introduction.

7.4   MLP NEIGHBOR MODELS

Neighbor information could be included in a standard MLP model by simply augmenting the spectral values for pixel i with the values for the pixels $n(i)$ in a fully–connected MLP. However, this would lead to an MLP with a large number of weights and would, moreover, violate assumption (7.2) that the label of a pixel depends on the spectral values of neighboring pixels only via their class labels. Such a model would in fact be using "texture" information to perform the classification and, as such, may cause confusion on class boundaries[3].

The architecture of the MLP can be modified to model the contextual information in the labels of neighboring pixels by augmenting a pixel with the spectral values of its neighbors. However, the appropriate model should have certain weights set to 0 and certain sets of weights held at the same value to avoid violating assumption (7.2).

To incorporate neighbor information into the MLP, we return to the theme of task-based MLPs considered in Chapter 6, p. 69. Each output unit of an MLP estimates $P(g^q|x)$ for some class g^q versus the other classes. Similarly, by construction, each hidden–layer unit in a task–based MLP can be considered to be discriminating between two classes or super-classes formed by the union of classes. If we designate a super class by S^j, then the jth hidden–layer unit estimates $P(S^j \mid x, \omega_h)$ versus the alternative $P(\overline{S^j} \mid x, \omega_h)$.

These estimates, $P(S \mid x)$, may be poor; for example, they may be small in magnitude across the entire range of the training data. However, the deficiencies in scale can be rectified by the Υ matrix of weights, which can be viewed as combining and re–scaling the probabilities of superclass membership to find $P(g^q \mid x)$ for the k classes of interest.

If, without loss of generality, we drop the j superscript and consider a particular hidden–layer unit, and take S_i to be the label of the ith pixel, then each S_i can be considered a binary random variable. An estimate of $P(S_i = 1 \mid x_i, \omega)$ is produced by the hidden–layer unit. If we consider the grid of these estimates $\mathbf{S} = \{P(S_i = 1), i \in \mathbf{I}\}$, then we can see that each hidden–layer unit produces a $[0, 1]$ grey–scale image. We refer to this as a "semi–classified image." These images were briefly discussed in Chapter 6, p. 69. See Figure 7.2 for a full set of semi-classified images from an MLP of size 6.6.8 used to classify Martin's farm.

We ignore the spectral information for the moment, and just consider the information contained in the labels of the neighboring pixels $n(i)$. For pixel i, we take

---

[3]See Bischof et al. (1992) for an example of such a classification scheme applied to remotely sensed images. While using texture information may cause confusion on the class boundaries, it may in some circumstances have compensating benefits.

Hidden layer Images for mlp.6.6.8

mlp.6.6.8

Figure 7.2 The hidden layer images from the standard MLP of size 6.6.8. Compare with Figure 7.7, p. 112.

each neighbor individually and estimate $P(S_{n(i)} = 1)$. Let

$$\gamma_{n(i)} = P(S_{n(i)} = 1 \mid \omega, x_{n(i)}) = \text{logistic}(\omega^T x_{n(i)}).$$

$\gamma_{n(i)}$ is then the output of a single–layer MLP of size $P.1$, referred to as a "neighbor sub–network". These neighbor sub–networks will be used in the construction of the neighbor MLP model. Assumption (7.2) resolves into

$$
\begin{aligned}
P(S_i|S_{r(i)}) &= P(S_i|S_{n(i)}) & (7.12) \\
&= c(S_{n(i)}) \exp\{\beta_0 + \beta_1 N(S,i)\} & (7.13) \\
&= \text{logistic}\{\beta_0 + \beta_1 N(S,i)\}. & (7.14)
\end{aligned}
$$

Using the arguments of Section 7.1.2.2 (p. 96), we model equation (7.12) as (7.13) where $c(S_{n(i)})$ is a normalizing constant, and $N(S,i)$ is the number of neighbors of i with label S. We then specialize this to (7.14).

We can estimate $N(S,i)$ by a sum of the $\gamma_{n(i)}$. For class label S and, say, four neighbors, we have four units indicating, by an output in the range of $[0,1]$, the probability that neighbor $n(i)$ has class label S. Then

$$P(S_i = 1|S_{n(i)}) = \text{logistic}\left[\beta_0 + \sum_{n(i)=1}^{N} \beta_{n(i)}\gamma_{n(i)}\right] \qquad (7.15)$$

$$= \text{logistic}[\beta^T \gamma],$$

where $\gamma = (1, \gamma_1, \ldots, \gamma_N)^T$. We call equation (7.15) a "neighbor network". It combines the four neighbor sub–networks to estimate the class of the ith pixel based on the estimates for the neighboring pixels. In order to integrate neighbor information into the MLP, we can use this sub–network in one of two different strategies, which are explored in Sections 7.4.1, 7.4.2 and 7.4.3.

7.4.1   Neighbor MLP model 1

The first option (model 1) is to add the neighbor sub–networks as units on the hidden layer. Then for each unit that estimates $P(S_i \mid x_i)$, there is one that estimates $P(S_i \mid S_{n(i)})$. We illustrate the modifications with a MLP of size 3.2.1 in Figure 7.3. The standard hidden layer consists of two units, labeled A and B. For a particular hidden–layer unit (say unit A) we do the following:

- replicate unit A (that is, the unit and the weights fanning into it) four times (we label these new units 1, 2, 3 and 4 in the figure);

- for each of these 4 units, we feed in the respective neighbors of pixel 0 so that:

    - spectral information from neighboring pixel 1 is fed into unit 1;
    - spectral information from neighboring pixel 2 is fed into unit 2;
    - spectral information from neighboring pixel 3 is fed into unit 3;
    - spectral information from neighboring pixel 4 is fed into unit 4.

    The outputs from units 1 to 4 then give the ascribed classification of the neighbors of pixel 0;

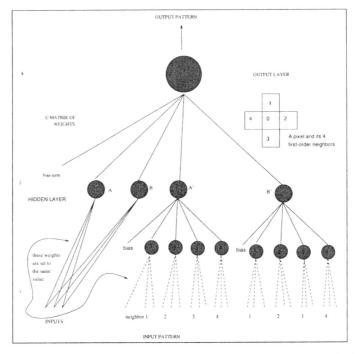

Figure 7.3   MLP neighbor model 1 adds neighbor sub–networks as units on the hidden layer. Then for each unit, A, that estimates $P(S_i \mid x_i)$, there is one, A', that estimates $P(S_i \mid S_{n(i)})$.

- the output from units 1, 2, 3 and 4 are combined in hidden–layer unit A'.

We preserve this correspondence between A and A' for the other hidden–layer units, so that the neighbor MLP now has twice the number of hidden–layer units as the original MLP. The MLP can be seen to consist of two parts at the level of the Ω matrix of weights:

1. a "spectral" part that uses the spectral information in pixel 0. This part of the network has h hidden–layer units and so can be described as a $P.H.Q$ network;

2. a "neighbor" part that uses the ascribed classification of the neighbors of pixel 0. This part of the network is not fully connected and so can not be described by the same simple scheme as the spectral network.

These two parts are combined at the level of the Υ matrix of weights. The model is then

$$z_q^* = P(g^q \mid x_{\bar{n}(i)})$$
$$= f_q(v_0 + v_1 y_1^* + \cdots + v_1'(y_1^*)' + \cdots) \qquad (7.16)$$

where

$$y_1^* = \text{logistic}(\omega_1^T x_i) \approx P(S_i^1 | x_i)$$

and

$$(y_1^*)' = \text{logistic}(\beta^T \gamma) \approx P(S_i^1 | S_{n(i)}).$$

Because of this architecture:

- the Υ matrix of weights is of dimension $q \times (2h + 1)$: and

- the Ω matrix is block–diagonal of size $2h \times (p + 1 + 5h)$, with a block of size $h \times (p + 1)$ followed by h blocks each of size 1×5.

7.4.2   Neighbor MLP model 2

Another option (model 2) is to feed the values $\gamma_{n(i)}$ as inputs, along with the spectral values for i, to the hidden unit assigned the task of separating S from $\bar{S}$. This means that the neighbor sub–networks are fanning directly into the hidden–layer units (see Figure 7.4).

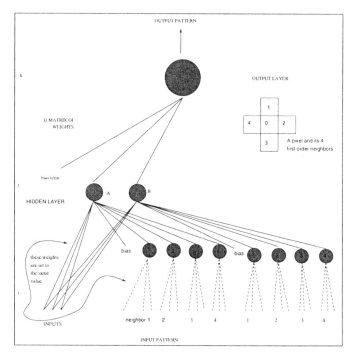

Figure 7.4   MLP neighbor model 2.

If we have a p-dimensional input space, h hidden–layer units and q output classes, this results in a MLP of size $(p + 1 + 5h).h.q$, assuming that the MLP is fully connected, and we have:

- the Υ matrix of weights is of dimension $q \times (h + 1)$, and;

- the Ω matrix is a block–matrix of size $h \times (p + 1 + 5h)$. It has a dense block of size $h \times (p + 1)$ adjacent to a block diagonal matrix of size $h \times 5h$, consisting of h blocks each of size 1×5.

Figure 7.4 shows a neighbor–based MLP (model 2) with three inputs, two hidden–layer units and one output.

The model is then

$$z_q^* = P(g^q \mid x_{\bar{n}(i)})$$
$$= f_q(v_0 + v_1 y_1^* + \cdots) \tag{7.17}$$

and

$$y_1^* = \text{logistic}(\omega_1^T x_i + \beta^T \gamma)$$
$$\approx P(S^1 | x_i, S_{n(i)}).$$

7.4.3 Neighbor MLP model 3

This model is similar to model 2 in that the neighbor information is directly processed by one of the existing units. It differs from model 2 in that this takes place at the output layer rather than the hidden layer.

The model is

$$z_q^* = P(g^q \mid x_{\bar{n}(i)})$$
$$= f_q \left(v_q^T y^* + \beta_{n(i)} \sum_{n(i)}^{N} z_q^*(x_{n(i)}) \right). \tag{7.18}$$

Implementing the model requires an additional N copies of the MLP. The easiest way to achieve this is to process the data, store the results, and then reprocess it using equation (7.18).

7.4.4 The boundary pixels

Consider classifying the pixels of a rectangular image of size $m \times n$. Now for pixels on the edge of this region, we do not have any neighbor information available. There are a number of ways to treat this problem. One attractive way is to consider the missing neighbors as "missing data". However, adapting MLPs to perform satisfactorily with missing observations is not trivial (Mathieson, 1997, gives a treatment of this problem).

There are a number of ways that we can attempt to supply the missing neighbor data. We can:

1. consider only
$$2 \leq i \leq m - 1$$

$$2 \leq j \leq n - 1,$$

so that we drop the edge pixels from consideration and only consider pixels for which we have neighbor information;

2. when considering an edge pixel, say pixel $(2, 1)$, make a neighbor for it by setting pixel $(2, 0)$ (the missing neighbor) to have the values of pixel $(2, 2)$;

3. consider the data as a torus so that we consider $(2, 0)$ as being $(2, n)$;

4. set $(2, 0)$ as $(2, 1)$.

Campbell and Kiiveri (1993) compare methods 1, 2 and 3 and conclude that the difference between them is minimal compared to the difference between a neighbor–corrected analysis and the usual analysis. We have opted for method 4 as it sets the value of the missing pixel to that of the closest observed pixel.

7.5   SEQUENTIAL UPDATING

The neighbor MLPs, models 1, 2 and 3, indicate how neighbor information can be incorporated into an MLP classifier. At the expense of increasing the necessary storage space, models 2 and 3 allow us to sequentially update the labels. This is done by introducing a discrete time step, indexed by m say, and iterating on the regularization process. Model 2 allows this to be done at the hidden layer and model 3 at the output layer. For model (7.16) is is not clear how a Hopfield network could be implemented. If it is done at the $(y_h^*)'$ semi–classified image then each S^j is updated solely on its neighbors and not its ascribed class.

Given an image $\mathbf{g}$, a model for $P(x|g_i)$ and $P(g_i|g_{n(i)})$, we can implement a Hopfield network. For example, using model (7.10) with $\alpha_i = 0$ we can calculate the distribution

$$P(g_i|x_i, g_{n(i)}) = \frac{1}{Z} P(g_i^q|x) \exp\{\beta N(g, i)\}, \tag{7.19}$$

where Z is a normalizing term, and $(g_i^q)^{(2)}$, the updated value for g_i is set accordingly.

There are several strategies for doing this:

- we can sample from the distribution $P(g_i|x_i, g_{n(i)})$. This is simulated annealing and allows less probable class labels to be selected (although they are selected with lower probability) and may prevent the process becoming stuck in undesirable local minima;

- we can set (g_i^q) to the mode of the distribution;

- we note that at both the hidden and output layers, we are starting with probabilities rather than class labels and hence we update these probabilities. While we do this in a naive way, it is justified (Saul and Jordan, 1996; Ripley, 1996, §8.4) under the name *mean field approximation*.

While for the first two options above it is clear that the algorithm terminates when a complete cycle through the data does not result in any pixel changing label, for the third option it is necessary to specify a stopping rule. This can be specified in terms of a minimal change, so that the algorithm terminates if $\sum_i ||z^*(x_i)]^{(m+1)} - [z^*(x_i)]^{(m)}|$ is less than some specified value, or by imposing a limit on the number of iterations.

7.5.1   At the output layer

For a cross–entropy MLP with a softmax activation function, since we have a set of estimates of the posterior probabilities of class membership,

$$z_q^* = P(g^q|x) \text{ for } q = 1, \dots, Q$$

we have Q images, each with a continuous class label.

An MRF model can then be formed as follows. We define

$$a_{q,i}^{(m-1)} = \exp\left\{\sum_{n(i)} [z_q^*(x_{n(i)})]^{(m-1)}\right\}$$

and then

$$[z_q^*(x_i)]^{(m)} = \frac{[z_q^*(x_i)]^{(0)} a_{k,i}^{(m-1)}}{\sum_{q_1} [z_{q_1}^*(x_i)]^{(0)} a_{q_1,i}^{(m-1)}}. \tag{7.20}$$

Assuming that f_q is a softmax function, ignoring the time step, and setting the βs to 1, it can be seen that models (7.18) and (7.20) are identical, both having the form

$$z_q^* = \frac{\exp\{v_q y^* + \sum_{n(i)} z_q^*(x_{n(i)})\}}{\sum_{q_1} \exp\{v_{q_1} y^* + \sum_{n(i)} z_{q_1}^*(x_{n(i)})\}}.$$

Thus model 3 (equation 7.18) offers a natural way of implementing a sequential updating scheme.

7.5.2   At the hidden layer

In a similar fashion, neighbor model 2 can be used to implement an Ising spin glass model at each hidden layer unit. Consider

$$y_h^* = f_h(\omega_h^T x + \beta\gamma)$$
$$= P(S_i^j | x_i, S_{n(i)}).$$

By introducing a time step, we can write

$$[y_h^*(x_i)]^{(m+1)} = P(S_i^j | x_i, (S_{n(i)})^{(m)}).$$

7.5.3   Combining information

While it is possible to train any of the models (including the neighbor–subnetworks) by any of the usual methods, it is not advisable to do so. All models involve combining information from two different sources to make a decision. Unfortunately, training the MLP on homogeneous regions means that both sources of information – the spectral values x_i and the class labels of $n(i)$ – will be giving essentially the same information. In this case, an unconstrained MLP may totally ignore one source of information. Because of this, it is necessary to impose some constraints on the MLP during the training process.

The way we have overcome these problems is to:

- train the Ω weights for the spectral data alone;

- set the weights for the neighbor–subnetwork to the same values as the Ω weights for the spectral connections;

- set the β weights to predetermined values.

Kiiveri and Campbell (1992) suggest that there is no advantage to using models more complex than (7.4) and in their examples take $\beta_0 = 0$ and $\beta_i = 1$. We set $\beta_1 = \beta_2 = \beta_3 = \beta_4 = -\frac{1}{2}\beta_0$ so that the network is summing the probabilities of the neighbors being in class S. If the sum is positive, the unit returns a "+" value (≥ 0.5), otherwise it returns a "-" value. Ripley (1986) suggests some simple geometric arguments for choosing a reasonable range of values for β.

Even with this approach, model 1 requires split training for the Υ matrix as both A and A' are estimating $P(S|.)$ and so the MLP may ignore one of them in training. In this case, the procedure we have used is to train one set of weights at a time (either the "spectral" or the "neighbor" weights) and then combine the weights to get the final MLP model. Because of this, and because model 1 does not lend itself in a natural way to the Hopfield extension discussed in Section 7.5, it is not as attractive a model as either 2 or 3.

7.6   EXAMPLE – MARTIN'S FARM

We reconsider the Martin's farm example from Section 6.4 (page 76). The example here differs in that the selected training sites are different, there are now 8 ground cover classes, and we use the ρ_c penalty term. The training classes and the number of pixels in each class are given in Table 7.1.

Table 7.1   The ground cover class and the number of pixels in the training data for each ground cover class for the Martin's farm example.

ground cover class	number of pixels
water	9
primary salt	96
secondary salt	67
remnant vegetation	54
wheat	78
lupins	66
medic pasture	72
dark green	168

The initial task-based MLP was of size 6.28.8 and was selected according to the algorithms described in Chapter 6 (p. 69). This was pruned down to a network of size 6.6.8, and as the pruning algorithm stopped here and an inspection of the eigenvalues of the Hessian matrix of the weights indicated that a minimum had been found in weight space, this was accepted as the final MLP. The neighbor modifications were then made to the network and a series of iterations performed on the image.

There is a marked improvement in the smoothness of the resulting image, but without any apparent loss of class delineation. Figure 7.2 gives the hidden layer outputs as semi-classified images. The weight simplification procedure discussed in Section 6.4.5 (p. 85) was applied giving the simplified matrix

$$
\Upsilon = \begin{pmatrix}
15 & 0 & 0 & 0 & -10 & -10 & 0 \\
0 & -20 & 0 & -20 & 15 & 0 & 0 \\
0 & 25 & 0 & 0 & -15 & 0 & 0 \\
0 & -15 & 20 & -10 & 0 & -10 & 0 \\
0 & 0 & 0 & 0 & 5 & 0 & 0 \\
-10 & 20 & 0 & 15 & 0 & -10 & -15 \\
0 & -10 & 0 & 0 & 0 & 10 & 0 \\
0 & 0 & 0 & 0 & 0 & 0 & 20
\end{pmatrix}.
$$

It can be seen that the ground cover classification "medic pasture" is formed by semi-classified image 5 minus semi-classified image 1.

Figure 7.5 shows the classification achieved with the standard MLP model. Figure 7.6 shows the classification achieved with neighbor model 3. Figure 7.7 gives the hidden layer outputs as semi-classified images as updated by neighbor model 2. In this example it was necessary to retrain the Υ matrix after applying neighbor model 2 to the hidden layer images as the proportion of the images classified as wheat shrunk markedly. Figure 7.8 shows the action of a neighbor MLP on a small section of the image.

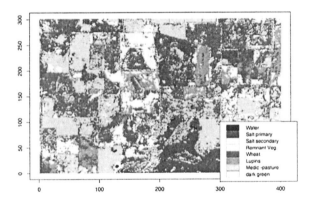

Figure 7.5   The classified image from the standard MLP of size 6.6.8 .

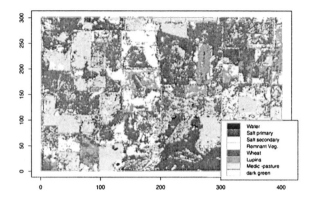

Figure 7.6   The classified image from the standard MLP of size 6.6.8 with the neighbor model 3.

7.7 CONCLUSION

We have shown how to modify the architecture of an MLP to incorporate neighbor information in line with assumption (7.2, p. 95) that the label of a pixel depends on the spectral values of neighboring pixels only via their class labels. In addition we have shown how to iteratively increase the probability of a label given the labels of the neighboring pixels. This is effectively an implementation of the algorithms of Geman and Geman (1984) and Besag (1986) in the context of an MLP model.

A number of substantial problems still remain. One is the question of how many iterations of the neighbor updating formulae to apply. An excess number of iterations tends to create structures – extending one class at the expense of others. This is not specific to the neighbor MLP models, it is a general problem for MRF models.

The other problem is the utility of counting neighbors in a certain class (or probabilities of neighbors). Dunne (1997) suggests that counting neighbors gives a very poor indication of the probability of a certain label. This study suggests that the prior induced by observing the neighbors of a pixel should be class specific. For example, in the remote sensing context, the prior induced by having 1 or 2 saline neighbors should be greater than the prior induced by having 1 or 2 crop neighbors.

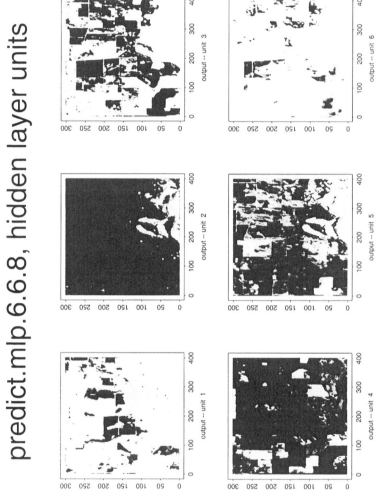

Figure 7.7 The hidden layer images from the standard MLP of size 6.6.8 updated with neighbor model 2. Compare with Figure 7.2.

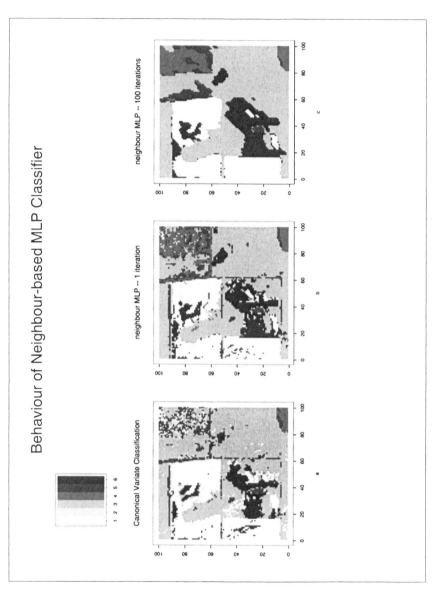

Figure 7.8 A region of Martin's farm as (a) displayed on the first canonical variate; (b) classified by the neighbor-MLP after one iteration; and (c) classified by the neighbor-MLP after 100 iterations. See Table 6.6 (p. 79) for a guide to the ground cover classes.

7.8   COMPLEMENTS AND EXERCISES

7.1   In Kiiveri and Campbell (1992) a simulated scene was constructed for use as a test case. The scene, shown in Figure 7.9, contains three classes (bush, wheat and pasture) and four spectral bands for each pixel. It was generated using covariance matrices and means (B_q and μ_q for $k \in \{\text{bush, wheat, pasture}\}$) typical of each class, with the exception that the means of the 3 classes were shifted towards their centroid to make the classification task harder. The scene was generated with a multivariate CAR model with $b_{10} = b_{01} = .3548$ (the horizontal and vertical parameter) and $b_{11} = -.1258$ (the oblique parameter). The scene is available as the data set `generated.image` in the `mlp` library.

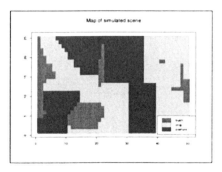

Figure 7.9   A map of the simulated scene from Section 7.1, p. 114.

Generated three 15×15 training areas (bush, wheat and pasture) using the same process. Let

$$\Sigma_q^{-1} = A \otimes A \otimes B_q,$$

where A is a 15×15 symmetric tridiagonal matrix with 1's on the main diagonal and $-.3548$ on the other two diagonals, B_q is the 4×4 covariance matrix for class k, and $\otimes$ is the Kronecker product[4]. If we take the Cholesky decomposition (Strang, 1988) of A and B_q so that

$$\begin{aligned} A &= LL^T \\ B_q &= KK^T \end{aligned}$$

then

$$\Sigma_q^{-1} = (L \otimes L \otimes K)(L \otimes L \otimes K)^T.$$

Hence we can find a square root of what will be a 900×900 matrix in this instance. Then take a vector X consisting of 900 observations drawn from a standard Gaussian distribution,

$$Y = \Sigma_q^{-1/2} X + \mu_q,$$

---

[4]If we have matrices $\underset{m \times n}{A}$ and $\underset{p \times q}{B}$ then the Kronecker product is

$$\underset{mp \times nq}{A \otimes B} = \begin{pmatrix} a_{11}B \dots a_{1n}B \\ \cdot \cdot \\ \cdot \cdot \\ a_{m1}B \dots a_{mn}B \end{pmatrix}$$

(where we cycle through the vector of means to make it the required length). The values μ_q are given in the script files.

This will have the required correlation structure. Reshape Y as a 225×4 matrix. This gives a 4–band data for a 15×15 rectangular region in "band interleaved by line" format.

Scale each band of the data to have a mean of 0 and variance of 1, and then scaled the image data with the same function. Figure 7.10 gives a plot of band 1 versus band 2 for the combined image and training data. It is clear that:

- it is an easy task to separate the three training classes;

- different decision boundaries, that give 100% accuracy on the training data, may give quite different results on the image data.

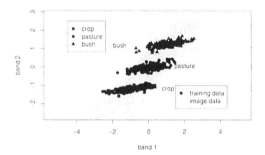

Figure 7.10   Band 1 versus band 2 for the simulated scene data (Section 7.1, p. 114).

7.2   First of all we consider the problem without using neighbor information. We consider several classifiers and also MLP models with different penalty terms and different values of λ. Table 7.2 gives the error rates for several classifiers. The MLP classifier was pruned from 4.3.3. to 4.2.3 by task-based pruning (Chapter 6, p. 69). The classified image produced by the MLP model is shown in Figure 7.11. The nearest neighbor classifier appears to give the best performance.

Tables 7.3, 7.4, and 7.5 compare several MLP classifiers with various values of a penalty term. Table 7.3 uses a weight decay penalty and Table 7.4 a fixed slope penalty[5]. It is apparent from an inspection of bivariate plots like Figure 7.10 that an MLP with no hidden layer (multiple logistic regression) could successfully classify these data. See Figure 7.5 for the results of doing this.

It is apparent from Tables 7.3, 7.4 and 7.5 that the penalized solutions are clearly better than the non-penalized solutions. For each of these tables the differences between the non-penalized and the penalized solutions are significant, while the differences between the penalized solutions are barely or non-significant. In addition it appears that a range of values of λ lead to good solutions.

The reader should add to this list by considering other classifiers such as:

- k nearest neighbor, for different values of k;

---

[5] This is a new penalty term introduced in Chapter 13, p. 219. It fixes the magnitudes of the weight vector fanning into a unit at a constant value.

- regularized discriminant analysis (Friedman, 1989). This is available in the R library **rda**.

7.3 Calculate the typicality indices (See Problem 3.9, p. 34) for the pixels of the simulated scene.

7.4 We now consider the neighbor information. Using the single–layer MLP, a comparison was made (given in Table 7.6) between the multi–color MRF using $P(g_i^q)$ and using g_i. It can be seen that the $P(g_i^q)$ MRF does much better, as would be expected as it uses more information than the g_i MRF. The classified image produced by the $P(g_i^q)$ MRF model is shown in Figure 7.12.

Apply a $P(g_i^q)$ MRF to the outputs of an MLP of size 2.2.3.

Table 7.2 A comparison of several classification techniques on the simulated scene. For each classifier the error rate on the image is given.

classifier	error rate
Linear discriminant functions	0.2536
Quadratic discriminant functions	0.2552
Nearest neighbor	0.2320
MLP (of size 4.2.3), a cross–entropy penalty function, softmax outputs	0.2600

Table 7.3 A comparison of an MLP (of size 4.2.3) with a cross–entropy penalty function, softmax outputs, and target values of 0 and 1, with various values of λ for a weight decay regularization term.

λ value	error rate
$\lambda = 0$	0.2600
$\lambda = 0.001$	0.2256
$\lambda = 0.01$	0.2264
$\lambda = 0.1$	0.2216
$\lambda = 1.0$	0.2232

Tables 7.8 and 7.7 give results for several MLP models using neighbor models 2 and 3 respectively. For row 3 of Table 7.8, as the outputs of the MLP are probabilities, model 3 is applied after model 2. The classified image produced by this model is shown in Figure 7.13. It can be seen that both models give a substantial improvement in accuracy over the standard MLP model.

Table 7.4   A comparison of an MLP (of size 4.2.3) with a fixed-slope training, a cross–entropy penalty function, softmax outputs, and target values of 0 and 1, with various values of γ. The magnitude of the weights fanning into a unit is fixed at γ.

γ value	error rate
unconstrained	0.2776
$\gamma = 6$	0.2208
$\gamma = 5$	0.2192
$\gamma = 4$	0.2224
$\gamma = 3$	0.2312

Table 7.5   A comparison of an MLP with a single layer, a cross–entropy penalty function, softmax outputs, and target values of 0 and 1, with various values of λ for a weight decay regularization term.

λ value	error rate
$\lambda = 0$	0.2256
$\lambda = 0.001$	0.2264
$\lambda = 0.01$	0.2192
$\lambda = 0.1$	0.2168
$\lambda = 1.0$	0.2096
$\lambda = 2.0$	0.2088

Table 7.6   A comparison of the error rates on the simulated scene using the MRF algorithm model (7.19) which uses the class labels g_i and neighbor model 3 which uses the probability estimates. It can be seen that neighbor model 3 is more accurate, presumably because it uses more of the available information. The first line of the table is the same model as the $\lambda = 2.0$ model in Table 7.5.

classifier	error rate
single layer MLP, $\lambda = 2.0$, probability MRF	0.0496
single layer MLP, $\lambda = 2.0$ class MRF	0.0728

Table 7.7   The initial and final error rates on the simulated scene using the neighbor model 3 (MRF algorithm equation (7.20)). The first line of the table is the same model as the $\lambda = 2.0$ model in Table 7.5.

classifier	initial rate	final rate
single layer MLP, $\lambda = 2.0$	0.2088	0.0496
MLP, cross–entropy outputs $\lambda = 0.1$	0.2216	0.0696
MLP, fixed slope, $\gamma = 6$	0.2208	0.0664

Table 7.8   The error rates on the simulated scene for the MLP models with neighbor model 2.

classifier	initial rate	final rate	model 3
MLP, logistic outputs, target values of 0 and 1, $\lambda = 0.1$	0.2216	0.0672	N/A
MLP, logistic outputs, target values of 0.1 and 0.9	0.2112	0.0704	N/A
MLP, cross–entropy outputs, $\lambda = 0.1$	0.2216	0.0664	0.0624

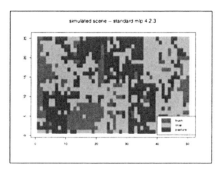

Figure 7.11   The simulated scene classified by a standard MLP of size 4.2.3

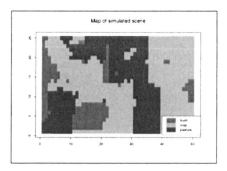

Figure 7.12   The simulated scene classified by a single layer MLP and a MRF.

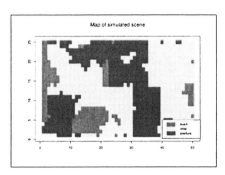

Figure 7.13   The simulated scene classified by neighbor model 2 with eight neighbors, and 20 iterations.

CHAPTER 8

INFLUENCE CURVES FOR THE MULTI–LAYER PERCEPTRON CLASSIFIER

8.1 INTRODUCTION

We consider the *influence curves* (IC) (Hampel et al., 1986; Huber, 1981) for the weights in an MLP model. These are described in the literature (Liu, 1994) only in the case of a regression MLP (so that $z^* = z$) for $P = 1$. Liu (1994) shows that in this case the *influence of the position in feature space* is bounded but that the influence of the residual is unbounded.

We describe the influence curves for the classification case for general P and for the penalty terms ρ_l and ρ_c. It is shown that the IC has a complex shape with influence curves exhibiting unbounded, bounded and redescending behavior depending on the architecture of the MLP, the penalty and activation functions and the dimension of the feature space. Among the conclusions it is shown that the MLP is not robust when $P > 1$ as some of the influence curves are unbounded. In addition it is shown that, for the ρ_c penalty function, the addition of a hidden layer causes unbounded influence curves to become bounded. The implications of this are explored further in Chapter 10, p. 159.

The influence curve is given graphically for the cases of two classes and one- and two-dimensional feature spaces. It is described and shown diagrammatically for higher dimensional feature spaces in Figures 8.4 (p. 133), 8.5 (p. 134) and 8.8 (p. 140).

These figures contain the essential conclusions of the chapter. The reader who is not interested in the derivations of the IC could simply refer to these figures.

8.2 ESTIMATORS

We briefly introduce some definitions and a statistical framework on which to base the discussion. The following terms are all descriptive of a statistical model:

parametric We assume that the model depends on a finite set of parameters. Fitting the model involves estimating these parameters;

distribution-free No distributional assumptions are made. For example, least squares is a distribution–free procedure. However, when the underlying distribution is assumed to be Gaussian, the least squares solution becomes a parametric estimator and is also the maximum likelihood estimator;

robust The term robust refers to methods of estimating parameters that are not sensitive to departures from the underlying distributional assumptions. (Hence **robust** methods are not **distribution-free** (Huber, 1981).) To quantify this, we impose a metric d on the space of distributions $\mathcal{F}$ and say that an estimator T is robust at a distribution F^0 if, for each $\epsilon > 0$, there exists $\delta > 0$ and $n_0 > 0$ such that for all F and $n \geq n_0$,

$$d(F^0, F) \leq \delta \Rightarrow d(L_{F^0}(T_n), L_F(T_n)) \leq \epsilon,$$

where $L_F(T_n)$ is the distribution of T_n if the sample is drawn from distribution F. Hence an estimator is robust if the mapping from distributions on the space of the random variable to distributions on the space of the estimates is continuous. An estimator has *robustness of efficiency* over a range of distributions if its mean squared error (or its variance for unbiased estimators) is close to the minimum for each distribution;

resistant A procedure is called resistant if it is insensitive to small changes in the sample, that is, to small changes in all the values, such as rounding, or large changes in a few of the values, such as gross errors and outliers.

While the terms "non-parametric" and "distribution-free" are used interchangeably in the literature, they are not the same. Using them interchangeably leads to the notational anomaly that fitting a non–parametric model then involves the estimation of parameters. Even more anomalous is the fact that often distribution–free models have a large number of parameters so that, if we adopt this nomenclature, we can heed the advice of Scott (1992, §2.4) on the relationship of "parametric" and "non-parametric" models,

> ... a nonparametric estimator should have many parameters, in fact perhaps an infinite number, or a number that diverges as a function of the sample size.

Staudte and Sheather (1990, §1.3.1) take a different approach that distinguishes between a "parameter" and an "estimand". For example, if we assume that F is a Gaussian distribution, then the mean and variance parameterize the distribution in that they determine F in the class of Gaussian distributions. The mean is thus a "parameter". If, however, we assume that F is a continuous distribution and calculate the median – this is an "estimand" that may characterize a quantity of interest but does not determine F in the class of continuous distributions.

Obviously, as both Scott (1992) and Ripley (1996) comment, the relationship between "parametric" and "non–parametric" estimators is not as clear cut as the terms would seem to imply. Ripley (1996, §2.2) comments that the more important distinction is between models that are constrained by having only a few parameters and those that are flexible enough to approximate a wide class of densities. The MLP model generally falls into the second class.

Hampel et al. (1986, p. 1–18) give a discussion of the relationship of robust to distribution-free[1] statistics, while Huber (1981, p. 7–9) discusses the relationship between robustness of efficiency and resistance. Hampel showed that the definition of robustness above is equivalent to the (weak) continuity[2] an estimator T; hence for all practical purposes, robustness and resistance may be considered to be synonymous. That is, if T is (weakly) continuous, then small changes in the data should only produce small changes in the estimate. If a set of training data contains no outliers or gross errors, then the robust estimator and the non–robust estimator should both give similar (and acceptable) estimates. However, in the case where the data contain aberrant observations, the robust estimator should still give an estimate that will generalize well.

It might seem unnecessarily formal to consider robustness in terms of weak continuity on $\mathcal{F}$. However it can be shown that ordinary continuity is not suitable as a robustness criterion. For example, consider the mean, which is not robust but does have the property that $T(x_1, \ldots, x_N)$ is ordinarily continuous as a function of the $x_1, \ldots, x_N$.

MLPs are in the same category as least squares linear regression in that they make no assumptions about the underlying error distribution. They simply minimize an objective error function, ρ, that is chosen as being sensible, and produce a mapping parameterized by a finite set of weights. Thus, while they are distribution–free, they are parametric procedures in the sense that a finite number of parameters are estimated during the fitting (or learning) process. Now if we make distributional assumptions, then the objective function ρ may be recovered via a likelihood equation. In this case our procedure, either linear regression or MLP modeling, may be derived as a maximum likelihood (ML) estimator. It is well known, for example, that least squares is not a robust method. However, the robustness or otherwise of the MLP is not immediately obvious; this question will be pursued in this chapter.

8.3   INFLUENCE CURVES

A brief discussion of robust statistics is given here. A full discussion is given in Hampel et al. (1986), Huber (1981) and Staudte and Sheather (1990). More introductory readings can be found in Goodall (1983) and Hogg (1979).

Following Huber (1981), a functional T on a space D is Gâteaux differentiable at F if there is a function a such that $\forall G \in D$,

$$\lim_{t \to 0} \frac{T\{(1-t)F + tG\} - T(F)}{t} = \int a(x)dG(x).$$

---

[1]Hampel calls this non-parametric.
[2]We say that "X_n converges to X in distribution" if $F_n(x) \to F(x)$ point–wise at every $x \in C(F)$, the set of continuity points of F. We write this as $X_n \xrightarrow{D} X$. This is termed convergence in measure, or *weak* convergence (Billingsley, 1995).

The left–hand side is the directional derivative of T at F in the direction of G. If we specialize D to the space of distributions, $\mathcal{F}$, and let $G = \delta(x)$ (the degenerate distribution with point mass 1 at x) we get

$$\lim_{t \to 0} \frac{T\{(1-t)F + t\delta(x)\} - T(F)}{t} = a(x). \qquad (8.1)$$

The quantity $a(x)$ can be interpreted as the scaled differential influence of one additional observation with value x as the sample size $N \to \infty$. Hampel has called it the "influence curve" (Hampel, 1974) and we write it as $\mathrm{IC}(x|F, T)$, showing the dependence on F and T.

8.4   M–ESTIMATORS

T is an M-estimator if its value is found as a solution to the minimization problem

$$T_N = \min_{\omega} \sum_{n=1}^{N} \rho(x_n; \omega).$$

T is indexed by N to indicate that this is an estimator derived from a sample of size N. Assuming that ρ is differentiable, we can alternatively solve the equation

$$\sum_{n=1}^{N} \psi(x_n, \omega) = 0,$$

where $\psi(x, \omega) = \dfrac{\partial}{\partial \omega} \rho(x, \omega)$.

Many standard estimators are M-estimators, including the sample mean which is the least squares estimator of location and has $\rho(x, \omega) = (x - \omega)^2$. Note that if $\rho(x, \omega) = -\log f(x; \omega)$, where f is a density function, we recover the usual Maximum Likelihood (ML) estimator, so that the M-estimators are extensions of ML estimators.

8.4.1   Influence curve of an M-estimator

Set

$$F^t = (1-t)F^0 + tF^1,$$

the model distribution F^0 contaminated by some other distribution F^1, and write

$$\begin{aligned} T' &= \left[\frac{d}{dt} T(F^t)\right]_{t=0} \\ &= \lim_{t \to 0} \frac{T(F^t) - T(F^t|_{t=0})}{t}, \end{aligned}$$

the ordinary derivative with respect to t, evaluated at $t = 0$. Let the contaminating distribution be $F^1 = \delta(x)$, then we recover the influence curve $\mathrm{IC}(x|F^0, T)$.

In particular, for an M-estimator, i.e. for the functional $T(F)$ defined by

$$\int \psi\{x; T(F)\} F(dx) = 0,$$

by inserting F^t for F and taking the derivative with respect to t, evaluated at $t = 0$, with $\psi'(x, \omega) = \frac{\partial}{\partial \omega} \psi(x, \omega)$, we obtain

$$\int \psi\{x; T(F^0)\} d(F^1 - F^0) + T' \int \psi'\{x; T(F^0)\} F^0(dx) = 0$$

or

$$T' = \frac{\int \psi(x; T(F^0)) d(F^1 - F^0)}{-\int \psi'(x; T(F^0)) F^0(dx)}.$$

After putting $F^1 = \delta(x)$, we obtain (by the properties of the δ function)

$$IC(x|F^0, T) = \frac{\psi(x; T(F^0))}{-\int \psi'(x; T(F^0)) F^0(dx)}. \tag{8.2}$$

Clarke (1983) justifies (8.2) in a general framework by proving the Fréchet differentiability of M-estimators even when ψ is not smooth. Writing (8.2) as IC $= k(\omega)\psi(x, \omega)$ we see that the influence curve of an M-estimator is proportional to ψ and moreover that the dependence on the distribution is subsumed into the proportionality constant k (Goodall, 1983).

8.4.2   Extension to the multivariate case

The IC is defined similarly for the multivariate case. In the case of M–estimators, we get

$$IC(x|T, F) = M(\psi, F)^{-1} \psi(x, T(F)),$$

where

$$M(\psi, F) = -\int \left[\frac{\partial}{\partial \omega} \psi(x, \omega) \right] dF(x),$$

an $R \times R$ matrix, where R is the total number of weights in the MLP model.

Hence we can see that the R–dimensional IC (where all R parameters may vary at once) is given by linear combinations of the one–dimensional ICs (where only one parameter is allowed to vary and the others are held constant).

8.4.3   Robustness of M–estimators

Let us specialize the discussion for a moment to location estimators

$$T_N = \min_{\omega} \sum_{n=1}^{N} \rho(x_n - \omega),$$

or equivalently

$$\sum_{n=1}^{N} \psi(x_n - \omega) = 0$$

$$\Rightarrow \sum_{n=1}^{N} \frac{\psi(x_n - \omega)}{x_n - \omega}(x_n - \omega) = 0$$

$$\Rightarrow \sum_{n=1}^{N} w_n(x_n - \omega) = 0.$$

Here T_N is a weighted sum of the discrepancies $d_n = x_n - \omega$ with

$$w_n = \frac{\psi(d_n)}{d_n}.$$

For regression, mean and variance estimation, the discrepancy is given by the residuals, whereas for multivariate problems the discrepancy is often the Mahalanobis distance. Now, as the discrepancy can be arbitrarily large, the influence of a single point may dominate the estimator T_N. A common method for making a procedure robust is to choose a function ψ and a cutoff point c so that:

$$\psi(d) = \begin{cases} d & |d| \le c; \\ < d & d > c; \\ > d & d < -c. \end{cases}$$

It is common in the literature on robustness to define an estimator in terms of the ψ function, as its behavior can be directly deduced from this function. There are two general approaches taken to this. We can:

1. set $\psi(x) = k$, a constant, for $|x| >$ some specified value – this gives the "Huber estimators"; or

2. make ψ "redescend" to 0, so that points far away have zero influence – this gives the "Hampel estimators".

Both estimators suffer from the problem of having to specify tuning constants; in addition, the Hampel estimators, while they are more resistant, may give multiple solutions to the minimization equation (Clarke, 1983).

For location estimators, ψ is generally chosen to be an "odd" function, so that $\psi(x) = -\psi(-x)$. In Figure 8.1, we show the following possible influence functions for a location estimator:

1. a least squares estimator – note that the IC is unbounded so that a single point may have an arbitrarily large influence;

2. a "Huber" estimator;

3. a redescending ("Hampel") estimator, using Tukey's biweight function

$$\psi(x) = x(a^2 - x^2)^2 I_{[-a,a]}(x)$$

(where I is the indicator function). This weight function leads to a widely used robust location estimator.

We note three features of ICs that are useful in understanding the behavior of the associated estimator. These are the:

1. *gross–error sensitivity.* This is the maximum effect a single observation can have on the estimator. It is the maximum absolute value of the influence curve, which is bounded for a robust estimator;

2. *local shift sensitivity.* This is the "worst effect of wiggling a point" and is (in the limit as the magnitude of the "wiggles" goes to 0) equal to the maximum

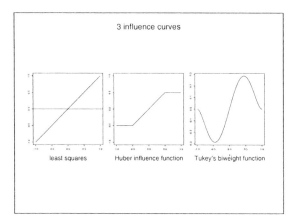

Figure 8.1   Three possible influence curves for a location estimator.

derivative of the IC. In the case where ψ is known to be continuous, the local shift sensitivity is guaranteed to be finite;

3. *rejection point,* which is the point beyond which observations have no influence. Only a redescending estimator has a finite rejection point.

Figure 8.2 shows the finite rejection point, the bounded gross error sensitivity and the finite local shift sensitivity for Tukey's biweight.

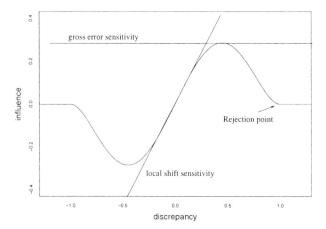

Figure 8.2   A re–descending (Hampel) influence curve showing the gross error sensitivity, the local shift sensitivity and the rejection point. This is Tukey's biweight function.

8.5   THE MLP

The standard MLP model with two layers of adjustable weights, P inputs, H hidden layer units and Q output units, and no skip– or intra–layer connections, was described in Chapter 2, p. 9. The MLP is an M–estimator as the weights are arrived at by minimizing an objective function, ρ. In this section we consider the *sum of squares* error function ρ_l with logistic output units. In Section 8.6 (p. 136) we discuss influence curves of the cross–entropy penalty function with the softmax activation function.

For a sample of size N and Q classes, we define ρ_l by

$$\rho_l(x, \omega) = \sum_{n=1}^{N} \sum_{q=1}^{Q} \frac{1}{2} \{t_{nq} - \mathrm{mlp}_q(x_n)\}^2,$$

where t_{nq} is the qth element in the nth target vector, and $\mathrm{mlp}_q(x_n)$ is the output of the qth output unit for a data vector x_n. For an MLP with one output, we have

$$\rho_l(x, \omega) = \sum_{n=1}^{N} \frac{1}{2} \{t_n - \mathrm{mlp}(x_n)\}^2,$$

and

$$\psi_{\omega_r}(x) = \frac{\partial}{\partial \omega_r} \rho(x, \omega)$$

$$= \sum_{n=1}^{N} \{t_n - \mathrm{mlp}(x_n)\} \frac{\partial}{\partial \omega_r} \mathrm{mlp}(x_n, \omega). \tag{8.3}$$

Equation (8.3) is a mapping from R^P to R (where we assume a given target value t, and the MLP has one output unit so that $Q = 1$). Note that it can be seen to be a product of two functions which, following Hampel et al. (1986), we call respectively, the *influence of the residual* and the *influence of the position in feature space*. As $\mathrm{mlp}(x)$ is bounded by 0 and 1 in the case of a classification task, the *influence of the residual* is also bounded by 0 and 1.

If the MLP is of size $P.H.Q.$, there will be $R = (P + 1)H + (H + 1)Q$ weights. There will then be QR ICs (one for each weight and exemplar target vector), and each IC will be a mapping from $R^P \rightarrow R$.

Note that in the following discussion, we only consider the one–dimensional ICs and the associated functions ψ_{ω_p}. We do not consider the multivariate IC; however, as we explained in Section 8.4.2 (p. 125), the multivariate IC can be shown to be a linear transformation of the vector of univariate ICs. Hence the shape of the univariate ICs will cast some light on the multivariate case, to the extent that if the univariate ICs are all bounded then the multivariate IC will also be bounded.

8.5.1   The perceptron

8.5.1.1   The case $P = 1$ We first try to get some indication of the general shape and properties of the IC by considering the simplest case, that of a single perceptron model and a one–dimensional feature vector. Assume that we have fitted a perceptron model to an arbitrarily large data set, with target values of $\{0, 1\}$, and that the parameter estimates are $\omega_0 = 0$ and $\omega_1 = 1$, so that our fitted model is

$$\begin{aligned} \text{mlp}(x) &= \frac{1}{1 + \exp(-\omega_0 - \omega_1 x)} \\ &= \frac{1}{1 + \exp(-x)}. \end{aligned}$$

Figure 8.3 shows a plot of the sigmoid function, which defines the fitted values for the perceptron, and the derivatives $\frac{\partial}{\partial \omega_0} \text{mlp}(x)$ and $\frac{\partial}{\partial \omega_1} \text{mlp}(x)$ which are, respectively, the influence curves for ω_0 and ω_1 with respect to position in feature space. We now take an individual point x, with label 1, and plot the one–dimensional influence curves (also shown in Figure 8.3). Note that the derivatives $\frac{\partial}{\partial \omega_0} \text{mlp}(x)$ and $\frac{\partial}{\partial \omega_1} \text{mlp}(x)$ are symmetric; however, when we multiply the *influence of the position in feature space* by the *influence of the residual*, the resulting function is no longer symmetric.

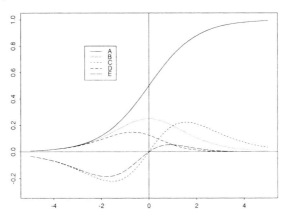

Figure 8.3   "A" is a sigmoid function. This is the fitted function of the simple perceptron. "B" is the derivative $\frac{\partial}{\partial w_0} \text{mlp}(w, F_n)$. "C" is the derivative $\frac{\partial}{\partial w_1} \text{mlp}(w, F_n)$. "D" is the $\psi_{w_0}(x)$ function and "E" is the $\psi_{w_1}(x)$ function.

We can readily see that the ICs for both parameters are redescending, and hence, for a one–dimensional feature space, the perceptron is a robust estimator (we show this analytically below). Both of the curves reach their maximum when the point x is misclassified, i.e., when the output of the sigmoid is less than 0.5. In Appendix B (p. 261), we extend this observation to larger, standard architecture, MLPs.

We note that the interpretability of these curves could be improved by re–parameterizing the weights of the MLP as $\text{mlp}(x) = [1 + \exp\{-a(b + x)\}]^{-1}$, so that we have a "scale" weight and a "location" weight. In Chapter 13 (p. 219) we take up the question of extending this re–parameterization to higher–dimensional feature spaces.

8.5.1.2 The case $P > 1$ We now move to the more complex case of a single perceptron model and a P-dimensional feature vector for $P > 1$. We fit the model

$$\text{mlp}(x) = \frac{1}{1 + \exp\{-(\omega_0 + \omega_1 x_1 + \omega_2 x_2 \ldots + \omega_P x_P)\}},$$

to a two–class problem, with the classes labeled by the target values 0 and 1. We consider the shape of ψ with one additional point from class 1. The ψ function for weight ω_p is

$$\begin{aligned}
\psi_{\omega_p}(x) &= \{1 - \text{mlp}(x)\}\frac{\partial}{\partial \omega_p}\text{mlp}(x, w) \\
&= \{1 - \text{mlp}(x)\}\frac{x_p \exp\{-(\omega^T x)\}}{[1 + \exp\{-(\omega^T x)\}]^2}.
\end{aligned}$$

The first thing we wish to establish about the IC is whether or not it is bounded. To do this, we examine the behavior of the ψ function as $|x| \to \pm\infty$ along two different sets of paths. We first interpret this limit by taking $x = \kappa$ where κ is a point on a line, ℓ, perpendicular to the decision hyperplane $\omega_0 + \omega_1 x_1 + \omega_2 x_2 \ldots = 0$. We then take the limits as $\kappa \to \pm\infty$. It is not necessary to determine an origin, or in which direction $\pm\infty$ lie; for our purposes we are only interested in the fact that there are limits in two different directions. We also consider the limit along paths that are parallel[3] to the decision hyperplane, so that the output of the MLP has a constant value. We write these two limits, respectively, as

$$\lim_{\substack{x=\kappa\to\pm\infty \\ \kappa\in\ell\,\parallel\,\omega^T x=0}} \quad \text{and} \quad \lim_{\substack{x=\kappa\to\pm\infty \\ \kappa\in\ell'\,\perp\,\omega^T x=0}}.$$

The *influence of the position in feature space*, for the weight ω_p, contains a term of the form

$$x_p\frac{\exp(w^T x)}{\{1 + \exp(w^T x)\}^2} = \frac{x_p}{\{\exp(-w^T x/2) + \exp(w^T x/2)\}^2}, \tag{8.4}$$

which will tend to 0 as $|\kappa| \to \pm\infty$ along any path perpendicular to $w^T x = 0$, due to the rapid growth of the exponential terms. This is essentially the same as the case of the one–dimensional feature space.

In addition, we can extend this result to any path that is not parallel to $\omega^T x = 0$. We consider (8.4), with $\omega^T x = \omega_0 x_0 + \omega_1 x_1 + \omega_2 x_2$ and, without loss of generality, we let $\omega_0 = 0$. We then consider $x = \kappa \in \ell$ where ℓ is a path that is not parallel to $\omega^T x = 0$, for example, the path $v_1 x_1 + v_2 x_2 = 0$. This implies that $x_1 = (-v_2 x_2)/v_1$, and then

$$\begin{aligned}
\omega^T x &= \omega_1 x_1 + \omega_2 x_2 \\
&= \omega_1(-v_2 x_2)/v_1 + \omega_2 x_2 \\
&= m x_2, \text{ say,}
\end{aligned}$$

---

[3]When dealing with dimensions higher than two, this is interpreted as a path parallel to some line contained in the decision hyperplane, which itself is a translate of a maximal (dimension $P - 1$) subspace of the feature space.

and (8.4) becomes

$$x_2 \frac{\exp(mx_2)}{\{1 + \exp(mx_2)\}^2}, \text{ which will } \to 0 \text{ as } |x_2| \to \infty.$$

However, when $|\kappa| \to \pm\infty$ along a path parallel to the decision boundary, $w^T x$ will take a constant value and the *influence of the residual* will also be constant. However, the *influence of the position in feature space* (8.4), will have a magnitude depending solely on the x_p term. This leads to the following observations:

1. as the x_p term for the bias is always 1, the *influence of the position in feature space* for the bias term will be a constant, k_1 say, as $|x| \to \infty$ along these paths;

2. if $w^T x = 0$ is perpendicular to the x_p axis, then the *influence of the position in feature space* for ω_p will also be a constant, k_2, along these paths;

3. for other weights, the influence of the position in feature space will be unbounded along these paths.

Bounds on the values of the constants k_1 and k_2 can be readily determined by considering the modulus of the expression in (8.4):

$$\left| \frac{\exp(w^T x)}{\{1 + \exp(w^T x)\}^2} \right| = \frac{1}{|\{\exp(-w^T x/2) + \exp(w^T x/2)\}^2|}$$

$$= \frac{1}{|\exp(-w^T x) + 2 + \exp(w^T x)|}$$

$$= \frac{1}{2|\cosh(w^T x) + 1|}$$

$$\leq \frac{1}{4}.$$

Hence the magnitude of the influence of the position in feature space is bounded, in cases 1 and 2 above, by $\frac{|x_p|}{4}$. This means that the ψ function for the bias term is bounded by $k_1 \leq 1/4$ as the bias input is always 1, while for case 2, the k_2 constant depends on the value of x_p, which is unbounded.

Hence we conclude that[4]

$$\lim_{\substack{x=\kappa\to\pm\infty \\ \kappa\in\ell \not\parallel w^T x=0}} \text{IC}(x|\omega_p) = 0 \tag{8.5}$$

and

$$\lim_{\substack{x=\kappa\to\pm\infty \\ \kappa\in\ell' \parallel w^T x=0}} \text{IC}(x|\omega_p) = \begin{cases} (a)\ k_1 \text{ where } |k_1| \leq m/4 \text{ for } \omega_0, \\ (b)\ k_2 \text{ where } |k_2| \leq \frac{|x_p|m}{4} \text{ for } \omega_p \\ \quad\ \text{if } w^T x \perp \text{ to the } x_p \text{ axis}, \\ (c)\ \text{grows without bound} \\ \quad\ \text{otherwise} \end{cases} \tag{8.6}$$

[4]The notation $\not\parallel$ stands for "not parallel to".

for some constants k_1 and k_2 which depend on the path ℓ. For the single perceptron $m = 1$ while for MLP models it will take other values (see below). We see that the IC for a simple perceptron displays a variety of behaviors and, except in the case $P = 1$, the single perceptron is not robust as the IC is unbounded.

8.5.2   Adding a hidden layer

We now look at an MLP with a hidden layer as discussed in Section 2.1 (p. 9). Without loss of generality, we still restrict the discussion to MLPs with one output. We can readily extend these results to an MLP with more output units by simply considering each output unit separately. Now considering $\dfrac{\partial \rho}{\partial \omega_{ph}}$ (equation (2.6)), we can see that the last term, $\dfrac{\partial f(y_h)}{\partial y_h} x_p$, is the ψ function for a single perceptron, and that this term is multiplied by a number of terms, all of which are bounded by 1 except for the $v_{q,h+1}$ term. Hence the asymptotic results of the previous section (equations (8.5) and (8.6)) all apply to the Ω matrix of weights for an MLP when we choose the constant

$$m = \|\Upsilon\| = \sum_{v \in \Upsilon} |v|.$$

These results (equations 8.5 and 8.6) are shown diagrammatically in Figure 8.4. We need now only consider the behavior of the weights in the Υ matrix.

8.5.2.1   MLP of size **1.1.1**   We consider the simplest case, an MLP of size 1.1.1, used to separate two classes. The Υ matrix consists of two weights, v_0 and v_1, and the Ω matrix also consists of two weights, ω_0 and ω_1. The fitted model is then

$$\text{mlp}(x) = \frac{1}{1 + \exp\left[-\left(v_0 + \dfrac{v_1}{1 + \exp\{-(\omega_0 + \omega_1 x)\}}\right)\right]}. \tag{8.7}$$

If we then take the limits of the product $\omega_1 x$ (so that we do not have to worry about the sign of ω_1), we see that

$$\lim_{\omega_1 x \to \infty} \text{mlp}(x) = \frac{1}{1 + \exp(-v_0 - v_1)} \tag{8.8}$$

and

$$\lim_{\omega_1 x \to -\infty} \text{mlp}(x) = \frac{1}{1 + \exp(-v_0)}. \tag{8.9}$$

The first thing that we note is that, unlike the single perceptron, the fitted values may approach any target values in the range $(0, 1)$ by varying the choice of v_0 and v_1. Hence, for the multi–layer perceptron, it does not appear unreasonable to have target values other than 0 and 1, and the common practice of using target values of 0.1 and 0.9 appears to have some vindication. In Chapter 9 (p. 143) we show that using target values of $\{0.1, 0.9\}$ may give a smoother separating boundary; however, it also has the undesirable effect that the outputs at the qth output unit may no longer be simply interpreted as an estimate of $P(C = k|x)$ (see Section 4.2, p. 35). More importantly for our discussion here, unless the values of the limits exactly

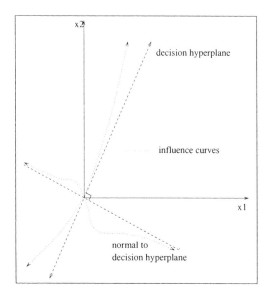

Figure 8.4   The shape of the ICs for the Ω weights for an MLP with a hidden layer. The IC, given as dotted lines, should be interpreted as a surface sitting over the page. In a direction not parallel to the decision hyperplane, the IC redescends to 0. In a direction parallel to the decision hyperplane, it is unbounded. There are two exceptions to this. One is the IC for the bias term, which is constant along paths parallel to the decision boundary. The other is the IC_{ω_p} when the decision boundary is perpendicular to the x_p axis. In this case also the IC takes a constant value along paths parallel to the decision boundary.

equal the target values, the *influence of the residual* will approach a constant in the range $(0,1)$, rather than going to 0.

The sigmoid function output by a single perceptron (Figure 8.3) is, after an appropriate affine transformation, an odd function; that is, $f(-x) = -f(x)$. However, the output of an MLP (with a hidden layer) does not have a comparable symmetry property. This can be easily seen from the fact that its limits as $x \to \pm\infty$ may take any value, as shown in (8.8) and (8.9). This means that the *influence of the residual*, (the $t - \mathrm{mlp}(x)$ term), will likewise not have this symmetry, and the ψ function for class 0 need not be symmetric with the ψ function for class 1.

For the second term of (2.5),

$$\frac{\partial}{\partial v_i}\mathrm{mlp}(x),$$

the *influence of the position in feature space*, we consider the asymptotic behavior. Taking limits and once again considering the variable $\omega_1 x$, we see that:

$$\lim_{\omega_1 x \to \infty} \frac{\partial}{\partial v_0}\mathrm{mlp}(x) = \frac{\exp(-v_0 - v_1)}{\{1 + \exp(-v_0 - v_1)\}^2},$$

$$\lim_{\omega_1 x \to -\infty} \frac{\partial}{\partial v_0}\mathrm{mlp}(x) = \frac{\exp(-v_0)}{\{1 + \exp(-v_0)\}^2},$$

$$\lim_{\omega_1 x \to \infty} \frac{\partial}{\partial v_1} \mathrm{mlp}(x) = \frac{\exp(-v_0 - v_1)}{\{1 + \exp(-v_0 - v_1)\}^2},$$

and

$$\lim_{\omega_1 x \to -\infty} \frac{\partial}{\partial v_1} \mathrm{mlp}(x) = 0.$$

Hence we can see that the ICs for the Υ matrix of weights are not redescending to 0, although they may be redescending to a constant value. We have coined the name *partially redescending* for this behavior. In addition, we can see that they may approach different limits in different directions.

8.5.2.2 MLP of size **P.1.1**   These results can be extended to an MLP of size P.1.1 by considering paths that are perpendicular to $w^T x = 0$ and paths that are parallel to it. The perpendicular paths are really the same as the one–dimensional case (i.e., $\kappa \in l \perp w^T x = 0$ gives a κ that is moving in a one–dimensional space). By considering a limiting value as $\kappa \to \pm\infty$, we see that the ψ_v function may approach non-zero limits in these directions. However, along paths parallel to $w^T x = 0$, the ψ functions will be constant. These results are shown diagrammatically in Figure 8.5.

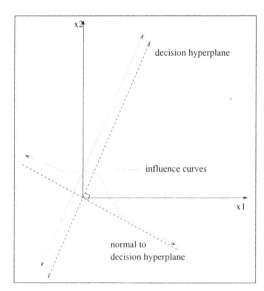

Figure 8.5   The shape of the ICs for the Υ weights for an MLP with a hidden layer. The IC, given as dotted lines, should be interpreted as a surface sitting over the page. For a non–bias weight, along a line not parallel to the decision hyperplane, the IC redescends to 0 in one direction and partially redescends in the other direction. For the bias term, the IC function partially redescends in both directions. The IC takes constant values along directions parallel to the decision hyperplane.

*8.5.2.3 MLP of size **P.H.**1*   We can also extend these results to an MLP of size
P.H.1. Let m be a vector of length $h + 1$ with $m_i \in \{0, 1\}$ and choose the m_i
such that $m_i = 1$ if the output from hidden layer unit i approaches 1, as $x \to \infty$,
and $m_i = 0$ if the output from hidden layer unit i approaches 0. Let $\bar{m}$ be the
complement of m, with the exception that for the bias term, m_0, $\bar{m}_0 = m_0 = 1$
always.

We now must consider $\kappa \in \ell$ such that $\ell \not\parallel w_j^T x = 0$ for j in $1, \ldots, h$, that is, ℓ
is not parallel to any of the decision hyperplanes associated with the hidden layer
units. Along any such path, x is moving into regions of feature space where the
hidden unit outputs are approaching either 0 or 1.

Then

$$\lim_{\substack{x=\kappa\to\infty \\ \kappa\in\ell}} \mathrm{mlp}(x) = \frac{1}{1 + \exp(-mv)}$$

and, traveling along the same line but in the opposite direction

$$\lim_{\substack{x=\kappa\to-\infty \\ \kappa\in\ell}} \mathrm{mlp}(-x) = \frac{1}{1 + \exp(-\bar{m}v)}.$$

We can also calculate the limits of the derivatives:

$$\lim_{\substack{x=\kappa\to\infty \\ \kappa\in\ell}} \frac{\partial}{\partial v_i} \mathrm{mlp}(x) = \frac{\exp(-mv)}{\{1 + \exp(-mv)\}^2} m_i \tag{8.10}$$

and

$$\lim_{\substack{x=\kappa\to-\infty \\ \kappa\in\ell}} \frac{\partial}{\partial v_i} \mathrm{mlp}(x) = \frac{\exp(-\bar{m}v)}{\{1 + \exp(-\bar{m}v)\}^2} \bar{m}_i. \tag{8.11}$$

So the MLP of size *P.H.*1 behaves in much the same way as the MLP of size
1.1.1 as $x = \kappa \to \pm\infty$. However, the behavior as x moves parallel to one or more
of the decision boundaries is harder to see.

8.5.3   Arbitrariness of re–descent direction

We have seen that for the Υ matrix of weights, the ICs will all be bounded, but
they do not necessarily redescend to 0. Observe, however, that as either m_i or $\bar{m}_i$
is equal to zero (for $i > 0$), either (8.10) or (8.11) will be equal to 0. Hence, moving
perpendicularly to the decision hyperplane, the IC will be redescending in either
one direction or the other.

We note, however, the arbitrary nature of this. If we take the MLP model (8.7)
and set

$$(\omega_0, \omega_1) \leftarrow -(\omega_0, \omega_1)$$
$$v_0 \leftarrow v_0 - v_1,$$

we obtain an MLP that produces exactly the same mapping. In this case though,
the ψ_{v_1} function redescends in the opposite direction.

8.5.4   A plot of the Υ and Ω ICs

Using the fact that the limit of a product of terms is the product of the limits of the separate terms (if all of the limits exist), and the formulae for the derivatives of the MLP with respect to weights on various layers, we can generalize these results to MLPs with any number of hidden layers. In addition we can see that skip–layer weights will behave like Ω weights if they are connected to the inputs, and like Υ weights if they are connecting hidden layers.

We illustrate our conclusions by taking an MLP of size 2.1.1 and treating it as though we have fitted it to an arbitrarily large data set and that our parameter estimates are

$$(\upsilon_0, \upsilon_1, \omega_0, \omega_1, \omega_2) = (-0.5, 1, 0, 1, -1)k.$$

We then calculate the ψ functions over the region $[-4, 4]^2$ and plot five functions $\psi_{\upsilon_0}, \psi_{\upsilon_1}, \psi_{\omega_0}, \psi_{\omega_1}$ and ψ_{ω_2} with $k = 2$ (Figure 8.6, p. 137). We have called the region around the decision boundary, the *region of active influence*. This region is characterized by an IC which has not approached its limit value more closely than some small constant. The name is chosen as "wiggling" a point in this region will change its influence whereas in the flat regions of the curve, "wiggling" a point has a negligible effect on its influence. As k is increased, the slopes of the curves will become steeper and the region of active influence will become narrower.

We see that the ICs for ω_0 and ω_1 are redescending while those for υ_0 and υ_1 are not. However, we note that, unlike the Huber IC depicted in Figure 8.1, they may reach a maximum and then partially redescend to a constant value (see the plot of ψ_{υ_1} in Figure 8.6).

In addition, we see (Figure 8.7, p. 138) that, as the weights are changed to rotate the decision hyperplane so that it is perpendicular to one of the axes (x_2 in this case), the shape of the IC changes. We see that for the weight w_2, instead of the influence growing without bound in the direction of the line $w^T x = 0$, it takes a constant value in that direction after the rotation has been applied.

8.6   INFLUENCE CURVES FOR ρ_c

We now consider the influence curves for the the cross–entropy penalty function,

$$\rho_c = -\sum_{q=1}^{Q} t_q \log(z_q^* / t_q),$$

described in Section 4.7, p. 43. This gives the *influence of the residual* as

$$\sum_{q=1}^{Q} \frac{-t_q}{z_q^*}, \tag{8.12}$$

which is unbounded as $z_q^* \to 0$. However, the cross-entropy penalty function is usually paired with the the "softmax" activation function

$$z_q^* = \frac{\exp(z_q)}{\sum_{q_1} \exp(z_{q_1})}$$

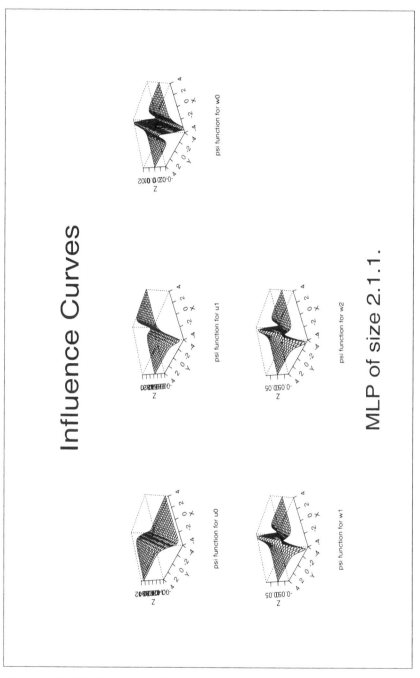

Figure 8.6   The five ψ curves for an MLP of size 2.1.1. The weight vector is ($\upsilon_0 = -0.5, \upsilon_1 = 1, \omega_0 = 0, \omega_1 = 1, \omega_2 = -1) * 2$. The factor of 2 is to make the shapes of the curves more pronounced. Note that the function ψ_{υ_1} redescends to a constant (not 0) in one direction. Note also the unboundedness of ψ_{ω_1} and ψ_{ω_2} along paths parallel to the decision boundary while ψ_{ω_0} is constant along these paths.

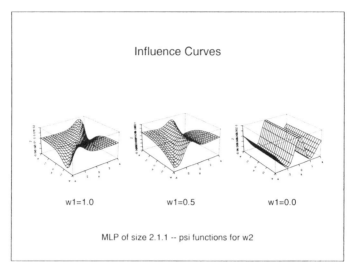

Figure 8.7   The same MLP as in Figure 8.6. We see the effect on ψ_{ω_2} of rotating the decision hyperplane until it is perpendicular to the x_2 axis.

at the output layer. When the derivatives are calculated,

$$\frac{\partial z_{q_1}^*}{\partial z_{q_2}} = \begin{cases} z_q^*(1 - z_q^*) & \text{when } q_1 = q_2 = q \\ -z_{q_1}^* z_{q_2}^* & \text{otherwise,} \end{cases}$$

we see that divisor of z_q^* can be canceled and so the combined influence curve will not be unbounded because of (8.12).

Here it is natural to consider combining the residual and position influences into a combined influence curve. The combined curve acts in many ways like a position influence curve in that it may still be unbounded in certain directions because of other terms.

For a single fixed input pattern we have, for $q = 1, \ldots, Q$ and $h = 1, \ldots, H + 1$

$$\frac{\partial \rho_c}{\partial \upsilon_{hq}} = (t_q - z_q^*) y_h^*$$

and for $h = 1, \ldots, H$ and $p = 1, \ldots, P + 1$ we have

$$\frac{\partial \rho_c}{\partial \omega_{ph}} = \sum_{q=1}^{Q} (t_q - z_q^*) \upsilon_{q,h+1} \frac{\partial f(y_h)}{\partial y_h} x_p.$$

Following this argument along the same lines as the argument in Section 8.5, we conclude that the behavior of $\text{IC}(x|\omega_p)$ is as described in equations (8.5) and (8.6) (p. 131). That is, it is unbounded in the general case along paths parallel to the decision boundary, $\omega^T x = 0$.

For the Υ weights we restrict our attention to an MLP of size $P.1.2$. We have $\mathrm{IC}(x|v_{hq}) = (t_q - z_q^*)y_h^*$ and thus

$$\lim_{\substack{x=\kappa\to\pm\infty\\\kappa\in\ell'\;\|w^Tx=0}} \mathrm{IC}(x|v) = \lim_{\substack{x=\kappa\\\kappa\in\ell'\;\|w^Tx=0}} \mathrm{IC}(x|v).$$

That is, the IC is constant along these paths as y^* and z^* are constant. Also

$$\lim_{\substack{x=\kappa\to\pm\infty\\\kappa\in\ell\;\|w^Tx=0}} \mathrm{IC}(x|v) = k \text{ where } 0 \le k \le 1$$

Note that as $y_h^* \to 0$ in one of these directions, (and $(z_q^* - t_q)$ may also approach 0) the IC will redescend in one direction.

However, when there is no hidden layer we have the logistic regression model and

$$\lim_{x\to\infty} \mathrm{IC}(x|\omega_{pq}) = (z_q^* - t_q)x_p.$$

Now:

- for the bias term ω_0 the IC is $(z_q^* - t_q)$ which is bounded above by 1 and below by 0, and constant along paths parallel to the decision boundary;

- along paths perpendicular to the x_p axis,

$$\mathrm{IC}(x|\omega_p) = (z_q^* - t_q)x_p,$$

  which is bounded by $|x_p|$;

- the IC is unbounded otherwise.

If $t \in \{0,1\}$ then $(t_q - z_q^*)$ will approach 0 in a direction perpendicular to the decision hyperplane $w^Tx = 0$. Thus, in the case where there in no hidden layer, there is a significant difference between the ρ_l and ρ_c (with softmax) MLP models.

This is illustrated in Figure 8.8 (which shows the IC for the logistic model) and Figure 8.4 (which shows the IC for the single layer ρ_l model). Here it can be seen that the ρ_l model will be more robust. This may account for the persistence in the literature of the ρ_l model for classification long after it became apparent that it was an inappropriate model.

Another point emerging from this section is that we might expect a $P.1.1$ model to be more robust than a $P.1$ logistic regression model. This is considered further in Chapter 10, p. 159.

8.7 SUMMARY AND CONCLUSION

After investigating the *influence curves* for the weights of a multi–layer perceptron classifier, our conclusions are briefly that:

1. the MLP can **not** be described as a robust estimator, due to the fact that the ICs for weights in the first layer are not all bounded for $P > 1$;

2. the ψ function (and the IC) for any weight on the first (input) layer of an MLP will behave as described in equations (8.5) and (8.6) (p. 131). That is, they display a mixture of redescending, bounded and unbounded behavior;

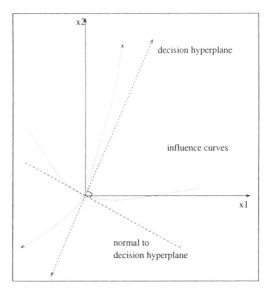

Figure 8.8   The shape of the ICs for a single-layer MLP with a ρ_c penalty function and a softmax activation function. The IC, given as dotted lines, should be interpreted as a surface sitting over the page. It can be seen that as a point moves from the origin in any direction its influence increases without bound. There are two exceptions: one is the IC for the bias term, which is constant along paths parallel to the decision boundary; and the other is IC_{ω_p} when the decision boundary is perpendicular to the x_p axis. In this case also the IC takes a constant value along paths parallel to the decision boundary.

3. for weights on the other layers, the IC will be bounded. Actually this can be seen without calculating the limits. The problem with the simple perceptron, and the first layer of the MLP, is the dependence on the magnitude of x. However, the outputs of the first layer are constrained to the interval $[0, 1]$ so that the inputs to the next layer are bounded (for given weights). Therefore, when the relevant ψ functions are calculated, they do not contain an unbounded term.

4. the unbounded ICs for the $P.H.Q$ MLP occur along paths parallel to the decision boundary. If we consider just a finite region containing the data (e.g., see Figure 8.6) then the regions of feature space with the highest influence lie close to the decision boundary. If a point in such a region is incorrectly classified, then in order to correctly classify it the decision boundary only has to be moved a relatively small distance.

5. we see a notable difference between the IC for a single layer MLP with:

   (a) a ρ_l penalty function; and

   (b) a ρ_c penalty function.

   In case (a) (as shown in equations (8.5) and (8.6)) the IC is bounded except along paths parallel to $\omega^T x = 0$; while in case (b) (except for the bias and

$\mathrm{IC}(x|\omega_i)$ along paths $\perp$ to the x_i axis) it is unbounded. The more widely unbounded nature of the ρ_c penalty function might lead us to expect less resistant behavior in some circumstances. We return to this point in Chapter 10 (p. 159). Figures 8.4, 8.5 and 8.8 illustrate the behavior.

The ICs for weights in the Υ matrix display a behavior that does not appear to be described in the literature. This behavior consists of redescending to a constant value, but not to zero, in directions perpendicular to the decision boundary. We have coined the name *partially redescending* for this behavior. This behavior means that the MLP estimator does not have a finite rejection point, so that points arbitrarily far from the decision boundary may still exert some appreciable influence. The implications of this for the resistance of the MLP model is still an open question.

CHAPTER 9

THE SENSITIVITY CURVES OF THE MLP CLASSIFIER

9.1 INTRODUCTION

We consider the question of the sensitivity of the MLP model to perturbations of a finite data set. We graphically demonstrating a number of facts about the MLP model, principally its sensitivity to a large number of conditions.

We fit MLP models to several small simulated data sets in one– and two–dimensional feature spaces, and examine the resulting models. The experiments cover the following cases:

1. two linearly separable classes in a one–dimensional feature space;

2. two linearly separable classes in a two–dimensional feature space;

3. two non–linearly separable classes in a two–dimensional feature space.

These experiments permit some simple qualitative assessments of the characteristics of the MLP model, and illustrate some aspects of its behavior. It is shown that:

- there are a large number of relevant conditions (starting weights, minimization routine, stopping criteria, encoding of target values, numerical precision). Due to the sensitivity of the MLP model and to the fact that the surface may have multiple minima, altering any one of these will very likely lead to a different solution;

- the MLP model may be extremely sensitive to a single outlying point. While this has been noted in the literature (Ripley, 1994a), we give a clear graphical demonstration of this fact;

- an MLP model with an excessive number of hidden layer units will not have any robustness property, and will readily overfit the data;

- there is a considerable difference between the influence curves (ICs) and the *sensitivity curves* (SCs). This is contrary to the experience with other models (Campbell, 1978) and is due to the iterative fitting procedure with a re-descending estimator and to the multiple minima of the error surface. We suggest that the shape of the SC is likely to prove as important as the shape of the IC in any question of robustness.

We also argue, on numerical evidence, that the beneficial effects of using initial weights with small magnitudes, and of early stopping, can be understood in terms of the shape of the IC.

9.2 THE SENSITIVITY CURVE

The IC is a major theoretical tool in the study of robustness and is also the basis of robustness; bound the IC and you have a robust estimator. However, the derivation of the IC depends on taking the limit as the sample size goes to infinity (Hampel et al., 1986, §2); hence we might expect some problems in using it as an applied tool when dealing with particular (finite) data sets. For example, the influence of one point may "mask" the influence of other points, so that the fitted model may have a sensitivity to the position of certain points that is quite different from that predicted by the IC.

While the gradient $\left.\dfrac{\partial\rho}{\partial\omega}\right|_{\omega_0}$ gives the direction of steepest descent in weight space at ω_0, it does not indicate how far to step in that direction to find a minimum, as this is also determined by the non-local shape of the surface. In addition, however far we should step in order to get to a minimum, the actual length of the step taken from the fitted model will also depend on the numerical minimization routine employed[1]. Hence even for a single step, the function $\partial\rho/\partial\omega$ does not reliably tell us how the parameters will change. Moreover, these same considerations will apply with each step of the iteration process.

For these two reasons, we consider a finite sample version of the IC, namely, the sensitivity curve (SC). If we have T_N, an estimator that takes a sample of size N, and a given sample $\{x_n\}_{n=1}^{N-1}$, we can plot $T_N(x_1, \ldots, x_{N-1}, x)$ as a function of x. Then, if we replace F by F_{N-1} and t by $1/N$ in the definition of the IC (equation 8.1, p. 124), we get the sensitivity curve

$$SC_{N-1}(x) = N[T_N(x_1, \ldots, x_{N-1}, x) - T_{N-1}(x_1, \ldots, x_{N-1})].$$

We then have three tools for examining the behavior of an MLP on a finite sample:

---

[1] A major source of variation, even between different implementation of the same numerical minimization routine, is the line search for the function minimum in the chosen direction. Minimization algorithms may use: an exact; an approximate; or no line search routine.

- the derivative $\left.\dfrac{\partial \rho}{\partial \omega}\right|_x$ can be plotted at the current model during the iterative fitting process – this is referred to as the "empirical IC";

- the decision boundary (where $mlp(x) = 0.5$) can be plotted at some point in the minimization process or for successive iterations;

- the SC can be calculated and plotted.

However, as we will see, each of these three methods (as well as the final model) are dependent on the starting values and the minimization routine. Hence we are only exploring a subset of the possible solutions to these three simple experiments.

9.3 SOME EXPERIMENTS

9.3.1 Experiment 1

We consider a small data set consisting of 10 points in $\mathbb{R}^1$, 5 points in each of 2 classes (Figure 9.1) and we fit an MLP model of size 1.1.1 (so that there are four weights).

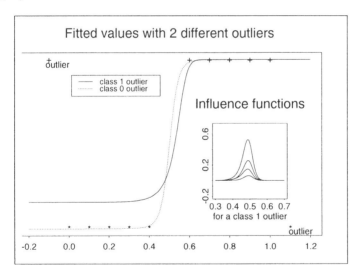

Figure 9.1   Two fitted curves with a 10–point data set and 1 additional outlying point of either class 0 or class 1. We can readily calculate the empirical IC for the 4 parameters under an additional observation of either class. See the sub–plot for the ICs for an observation of class 1.

We add an additional point, an outlier, and refit the model from the same starting weights, to see what effect this has on the fitted model. When we add, respectively, a point from class 1 and a point from class 0, we get a very different outcome, both shown in Figure 9.1. In this instance, the asymmetric nature of the ICs leads to noticeably different results in the two cases[2]. Note that the only effect that the outlier has is to change the shape of the fitted sigmoid (as shown in Figure 9.1), and that the decision boundary remains at approximately the same point and the classification accuracy remains the same.

The inset plot in Figure 9.1 shows the empirical IC functions for an observation of class 1, while Figure 9.2 shows the same but with target values of 0.9 and 0.1. Note that in Figure 9.2 the shape has changed considerably and that it is now clearly discernible that the ψs for v_1 and v_2 redescend to 0 in one direction and to a constant value in the other direction. This is described as a "partially redescending" IC function and was discussed in Section 8.7 (p. 139).

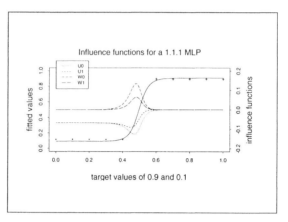

Figure 9.2   The empirical ICs for the four weights. Note that the ICs for weights w_0 and w_1 are partially redescending.

After one iteration the empirical IC and the SC are very similar in shape. However after repeated iterations the shapes of the two curves diverge due to the iterative fitting procedure. Figure 9.3 gives the shape of the SCs after 2000 iterations. Note that despite its more complex shape we still see a central *region of active influence* (Section 8.5.4, p. 136) and then convergence to a constant (perhaps zero) as $x \to \pm\infty$.

9.3.2   Experiment 2

We consider a simple problem with two classes and two features. The classes are linearly separable and were chosen from uniform distributions on rectangular regions (Figure 9.4). The procedure of the experiment was to progressively move one point, x, and note the effect on the separating hyperplanes. For each movement of the point, the MLP was refitted using the same starting weights.

For comparison with the MLP classifier, we also calculated Fisher's linear discriminant function (Chapter 3, p. 19). As the LDF gives a hyperplane onto which the data are projected, we plot the hyperplane given by the normal vector to the LDF and passing through the grand mean of the classes. Clearly both the LDF

---

[2]The same starting values were used in each case.

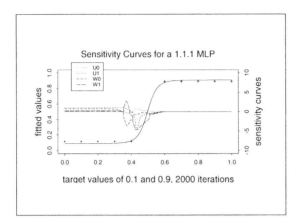

Figure 9.3   The sensitivity curves for the 4 weights $\{v_0, v_1, \omega_0, \omega_1\}$ calculated after 2000 iterations. Note that the iterative nature of the MLP means that the sensitivity curve presents a more complex picture than the empirical IC function.

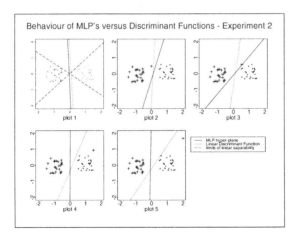

Figure 9.4   The MLP and LDF decision boundaries with one outlying point. In plot 3 the MLP follows the aberrant point whereas in plot 5 it is quite insensitive to it. The LDF is increasingly influenced by the point x as its Mahalanobis distance is increased. Plot 1 also shows the limits of linear separability.

hyperplane and the MLP separate the two classes (Figure 9.4 plot 1). Note that LDFs are not robust and that, as a point x is moved, its influence on the LDF is proportional to its distance from its class mean, which can be made arbitrarily large (e.g., see Figure 9.4 plot 5).

Like Experiment 1, the MLP is small enough (of size 2.1.1, with 5 adjustable weights)[3] to investigate its actions in some detail.

In plot 1 (Figure 9.4), the limits of linear separability are shown. As we shall see, these are quite important in describing the behavior of the MLP, whereas for the LDF the question of the separability of the classes is not relevant.

It is clear that the MLP does have some robustness property, as evidenced by the fact that in plot 3 the MLP follows the aberrant point, whereas in plots 4 and 5, where the classes are not linearly separable, it no longer does so. We note that the final MLP models in plots 4 and 5 are very similar to that in plot 1.

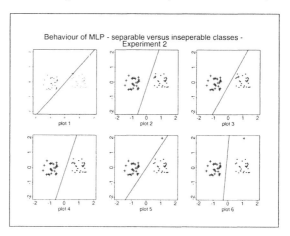

Figure 9.5   The final decision boundary for the data from Experiment 2. As we move the point x we can see the resulting change in the fitted model. Note that the MLP is insensitive to the point x when it is masked by other data points (plots 2, 3 and 4). Note also the large difference in the decision boundary between plots 5 and 6, for a very small movement in the point x.

We look more closely (Figure 9.5) at what happens when the point x is moved across the boundary between separability and non–separability. The MLP separates the two classes whenever it can, even when this leads to what might be considered an unsatisfactory fitted model (Figure 9.5 plot 1 and plot 5). However, when it is no longer possible to produce linear separation (as in Figure 9.5, plots 2, 3 and 4), we see that the influence of the moving point is markedly reduced as it is "masked" by other points.

In the plot of the SC for ω_2 (Figure 9.6), we see that it has a distinctive ridge that follows the line of linear separability and gradually decays. Any failure to separate two linearly separable classes in practice is due to the limitations of the function

---

[3]The MLPs were always started from the same random weights, have the maximum iterations set to 5000 and the tolerance set to 10^{-6}. Unless otherwise specified, a scaled conjugate gradient minimization routine was used (see the software documentation for further details).

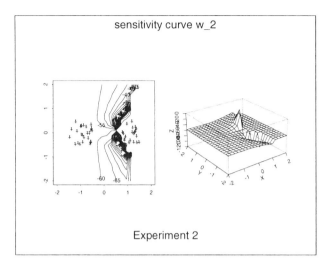

Figure 9.6   Contour and perspective plots for the sensitivity curve for the weight ω_2.

minimization routine, the stopping criterion, or the precision of the computations. The fact that the classes are not separated in Figure 9.5 plot 6 was due to the fact that the model was fitted using single–precision arithmetic. When the model was refitted using double precision, the two classes were separated, as expected[4].

In Figure 9.7 we see what happens when a second hidden-layer unit is added. Clearly any robustness property that the MLP model might have is lost when the model is over-parameterized.

9.3.3   Experiment 3

We consider a data set consisting of two classes that are not linearly separable (Figure 9.8) and note a number of things about the fitted MLP model. As the IC and the SC are quite complex in this case, we start by plotting the decision boundary.

We consider three different cases:

case 1 an MLP of size 2.3.1, with 2 different starting values (Figure 9.8, plots 1 and 2);

case 2 an MLP of size 2.6.1, so that there are more hidden–layer units than are necessary to separate the classes (Figure 9.8, plot 3);

case 3 an MLP of size 2.6.1, with one point moved over into the region of the other class (ie. one point is misclassified) (Figure 9.8, plots 4 and 5).

---

[4] All the fitted models reported here were fitted using double precision arithmetic. This was the only model fitted using single precision. The single precision code is no longer supported. However, it is very interesting that a change in precision can affect a statistical model so markedly. See Section 10.3.1, p. 163, for another example of this.

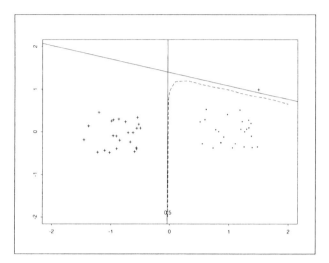

Figure 9.7   The decision boundary and the separating hyper–planes are plotted for an MLP model of size 2.2.1, fitted to the illustrated data set. Clearly when the MLP model is over–parameterized, any robustness property is lost.

9.3.3.1 Case 1, an MLP of size 2.3.1   (Figure 9.8, plots 1 and 2.) We observe two things about case 1. The first is that the decision region is not bounded by straight lines. The non–linearity of the sigmoid function means that where two hyperplanes intersect, the decision region may be bounded by a continuous curve (Lui, 1990). The other is that different starting values may give rise to very different final models.

9.3.3.2 Case 2, an MLP of size 2.6.1   (Figure 9.8, plot 3.) As only three of the six hyperplanes are necessary to separate the two classes, we may conclude that three of the six are redundant. The redundant hyperplanes may behave in the following ways:

1. they may have a tendency to lie on top of each other and thus perform the same task (Figure 9.8, plot 3);

2. they may lie outside the region containing the classes and thus return the same value for all classes (their output then becomes incorporated into the bias terms for the units on the next layer);

3. they may intersect the classes but have low weights fanning out (or into) them, thus contributing little (or nothing) to the overall fit[5];

4. the "redundant" units may contribute to forming the boundary of the decision region.

By altering the starting values, it would seem that case 1 is the most common.

---

[5]See Section 5.3.4 for a discussion of a test proposed by White (1989), which is based on the expectation that this will be the behavior of redundant hyperplanes.

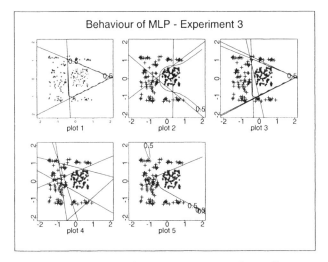

Figure 9.8   Plots 1 and 2 show different solutions arising from differing starting values. The straight lines are the rows of the Ω matrix and the curved lines are the decision boundary where mlp$(x) = 0.5$. Plot 3 shows the result of fitting a model with excess hidden layer units. Plot 4 shows the decision hyperplanes formed by the first–layer weights when one point in the training data is mis-specified. The mis-specified point is at approximately $(-1, 0)$. Plot 5 shows the resulting decision boundary and demonstrates that an MLP with one hidden layer can form disjoint decision regions.

9.3.3.3 Case 3, an MLP of size 2.6.1, with one point incorrectly allocated   (Figure 9.8, plots 4 and 5.) In case 3, the MLP has formed a disjoint decision region so as to be able to capture the single outlying point (Figure 9.8, plot 4 (Ω weights) and plot 5 (decision boundary)). Lippmann (1987) stated that the 2-layer MLP forms decision regions that are polyhedra or the complement of polyhedra in $\mathbb{R}^P$. Gibson and Cowan (1990) give some history and references on the development of an understanding of the decision regions of an MLP model.

How we view the behavior of the model depends on how we view the point x. If x is a "good" data point, then the MLP has successfully found a decision boundary that separates the training data, and the decision boundary will presumably also perform well on a new set of data drawn from the same distribution. On the other hand, if we view x as an outlying point, then the MLP has overfitted the training data. This overfitting could have been prevented by restricting the number of hidden–layer units to three (see Section 9.4.4, p. 154).

9.4   DISCUSSION

9.4.1   The shape of the sensitivity curves – sensitivity peaks

As we have noted, single points (or small regions in the feature space) may be associated with very large influences. This can occur both in the case where the model is over–parameterized and where the model is correctly parameterized.

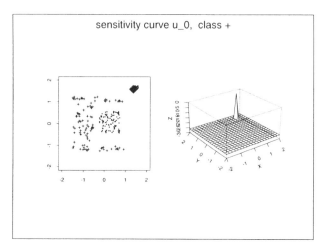

Figure 9.9   Contour and perspective plots for the sensitivity curve for v_0 for a point of class "+". The plot shows a pronounced "influence peak" at one point (approximately $(1.5, 1.5)$), to the extent that all other features are obliterated.

For the first case, see the fitted surface of Figure 9.8 plot 6, where the single point x has a great effect on the fitted model, while for the second case, see Figure 9.9 where we plot the SC for case 1 (Experiment 3). This plot shows a pronounced sensitivity peak, suggesting that a point at this position in feature space will have a major influence on the fitted values. To test this we fit an MLP model with, and without, a point at the position of the sensitivity peak. The difference in the fitted models is quite clear[6] in Figure 9.10.

9.4.2   Partially redescending ψ functions for Υ weights and the shape of the fitted surface

Following the heuristic argument of Lippmann (1987), it can be argued that the position of the decision boundary is determined, in the first instance, by the Ω weights, and that the Υ matrix combines the first–layer decision hyperplanes to form curved decision regions and determines the shape of the fitted surface $z^*(x)$. However, as we have shown previously, the IC for the Υ weights may be "partially redescending" and redescend to a constant along paths orthogonal to $\omega^T x = 0$. This means that the MLP estimator does not have a finite rejection point along these paths. (See the plot of the SCs for Experiment 2, where we can see that the SCs for the Υ weights have a constant, non-zero, value for a misclassified point of class "+", whereas the SCs for the Ω weights are essentially zero.)

Because of the difference in the behaviors for the two layers of weights, a point x may have an influence on the shape of the fitted surface without appreciably moving the decision boundary, as shown in Figure 9.1, p. 145. Hence, the decision

---

[6]Note the position and existence of the spike may be quite dependent on the initial values and the minimization routine.

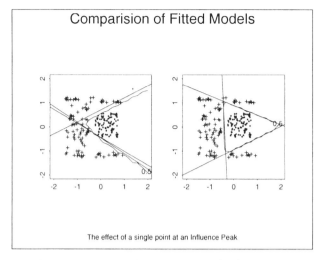

Figure 9.10   A comparison of the fitted models with *(left)* and without *(right)* a single additional point at the influence peak.

boundary may be more robust than the posterior probability estimates. A similar point about the resistance of the probability estimates versus the resistance of the decision boundary (in a more general setting) is made in Friedman (1996).

9.4.3   The changing shape of the empirical influence curve through time

We suggest that a consideration of the *region of active influence* can show why it is an essential part of the MLP methodology to start with weights chosen in a small interval around 0, and the reason for the effectiveness of "early stopping" (given that we have used initial weights of small magnitude).

Consider the arguments for the effectiveness of early stopping[7]. They are roughly that training should be stopped after the MLP has modeled the structure in the data and before it has started to model the noise. However, this begs the question of why the MLP should model the structure first.

Now, if we take it as being the case that the greatest reduction in ρ is effected by getting the structure right, as it will be in any non–pathological example, then the question becomes, why should the direction of steepest descent from the starting weights be in the direction that will model the structure of the data rather than the local noise?

The answer, we suggest, is that when the weights are initially of small magnitude, the *region of active influence* will be wide (perhaps as wide as the data) and the local shift sensitivity low. Hence all points in the data set will exert a roughly equal influence and modeling the structure will lead to a more rapid decrease in the penalty function than modeling the noise.

---

[7]See the critical comments on early stopping by B. Ripley in the discussion of Cheng and Titterington (1994).

Conversely, if the initial weights are large then only points very close to the decision boundary will have an active influence. In this situation it is often the case that the initial weights are near to an undesirable local minima and the MLP decision boundary will only move a short distance to accommodate points close to the decision boundary.

In support of this, we consider the evolution of the separating hyperplane through time. Figure 9.11 shows the data set from Figure 9.4 plot 3 and the hyperplane after various iterations. We note that the hyperplane moves to a position (lines 4 and 5) that we might consider "optimal" quite rapidly, and then moves towards its final position. In this example, early stopping might have selected the optimal decision boundary. Looking at the outputs of the MLP over the region of feature space containing the two classes (Figure 9.12), we note that as the MLP moves to a position where the hyperplane separates the data, it also increases the length of the vector of weights, which has the effect of making the sigmoid "steeper" and thus getting closer to the target values for the two classes. It is only after the sigmoid has become sufficiently steep that it can move to accommodate the point x. As the sigmoid is getting steeper and the length of the weight vector is increasing the *region of active influence* is becoming narrower.

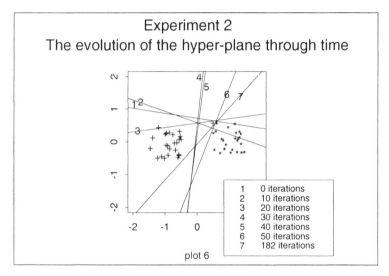

Figure 9.11 We track the evolution of the decision hyperplane through time. The hyperplane moves to an optimal point quite early (40 iterations, with scaled conjugate gradients) and then follows the aberrant point x.

9.4.4 Excess hidden layer units – no robustness property

As is obvious from Figure 9.8 plot 5, when a model with an excess number of hidden layer units is fitted, any robustness property that the MLP has breaks down. Unfortunately it is only for low–dimensional problems, such as the ones considered here, that we can see the number of required hidden layer units.

Comparing the fitted models in Figure 9.8 plot 1 and plot 5, we see that the fitted surface in plot 5 has a spike at the point x. It also has a decision boundary with a higher curvature, and the magnitude of the weights is much greater. Many attempts have been made in the literature to prevent this non-robust behavior by limiting the curvature of the fitted surface or by limiting the magnitude of the weights. See Section 5.4 (p. 62) for a brief survey of the literature on this question.

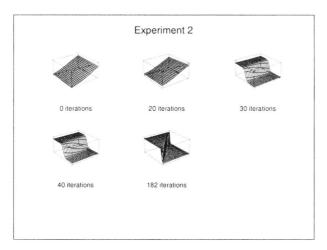

Figure 9.12   The fitted surface mlp(x) for various iterations. It is only after the sigmoid has become sufficiently steep that it can track the aberrant point x.

9.4.5   Target values of 0.1 and 0.9

With a sigmoid activation function, the output is restricted to the range $(0, 1)$; however, setting target values of 0 and 1 may cause the weights to grow extremely large during the fitting procedure. Pao (1989) suggests the practice of setting the target values to 0.1 and 0.9. Sarle (1996) calls this practice a "gimmick" and says that it probably derives from the slowness of standard "backpropagation". While this may be true, we have found that $\{0, 1\}$ targets tend to give a decision boundary that is closer in appearance to that of a hard–delimiter activation function network, whereas with targets of $\{0.1, 0.9\}$, the decision boundary has a smoother appearance; this is shown in Figure 9.13 where both target encodings are used.

This encoding could be viewed as a naive attempt at penalized fitting. However, it should not be used in applications where the outputs of the MLP are interpreted as estimates of the posterior probability of class membership (Section 4.2, p. 35).

9.4.6   The question of uniqueness of the weights

Some linear transformations of the weights and reorderings of the hidden–layer units will leave the solution unchanged. Bishop (1995a, §4.4) notes that an MLP

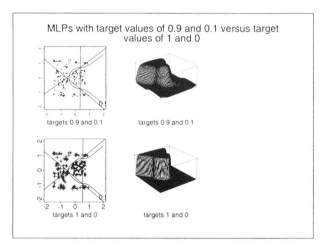

Figure 9.13   In the upper pair of plots we see the result of using target values of 0.1 and 0.9. The lower pair of plots show the outcome of using target values of 0 and 1. It can be seen that the decision boundary in the upper plots is smoother than that in the lower plots.

with tanh activation functions[8] and h hidden–layer units will have $h!2^h$ symmetric solutions. One theoretically satisfying resolution to this is to take a weighted sum of the $P(C_q|x, \omega)$ for each local minimum (Ripley, 1996, §5.5). Unfortunately it is often a very CPU–intensive task to find a single local minimum.

In addition, the unboundedness of the weights may lead to an infinite number of solutions. For example, take the data set from Experiment 2. As the magnitudes of the weights approach ∞, the output of the MLP will approach a Heaviside step function. However, as any step function that lies between the two classes will give rise to a perfect classification, there are an infinite number of possible solutions, all equivalent in that they involve a decision hyperplane drawn between the classes, and all involving infinite weights. This is a parameterization problem and it is only the parameter estimates that are unreasonable, the posterior probability estimates are quite sensible.

9.4.7   Functions of the weights

If we let $\zeta(\omega)$ be a function of the weights $\omega = \text{vec}(\Upsilon^\tau, \Omega^\tau)$, then

$$\frac{\partial \zeta\{\omega(x)\}}{\partial x} = \left(\frac{\partial \zeta}{\partial \omega}\right)^\tau \left(\frac{\partial \omega}{\partial x}\right).$$

By this method we can examine functions of the weights such as the slope and intercept of the decision hyperplane. If we take $\zeta(\omega) = \text{mlp}(\omega)$ we have

$$\frac{\partial \text{mlp}\{\omega(x)\}}{\partial x} = \left(\frac{\partial \text{mlp}}{\partial \omega}\right)^\tau \left(\frac{\partial \omega}{\partial x}\right)$$

---

[8]The tanh function is chosen as it is an odd function and makes the calculation easier. Corresponding results hold for the sigmoid activation function.

and, if we approximate $\dfrac{\partial \omega}{\partial x}$ by the sensitivity curve SC_ω, we can summarize the sensitivity of the MLP model in one function. (See Figure 9.14, where such a combined SC is given for the data set from Experiment 2). Note that this is the sensitivity of the fitted surface, rather than the sensitivity of the parameters of the MLP.

To do this we need a good estimate of the SC. The estimates that we have used here are all from a single starting point and could be improved by averaging over a number of starting points.

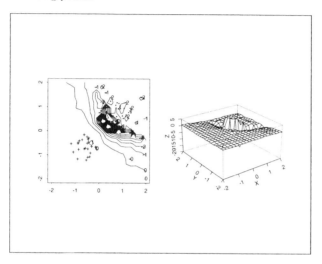

Figure 9.14   The sensitivity curve of the fitted surface of the MLP.

9.5 CONCLUSION

9.5.1 SCs and ICs

Due to the iterative nature of the learning process, there may be a considerable difference between the (empirical) ICs and the SCs. However, if the fitting procedure is stopped after one iteration, the agreement between the curves is much better. For example, compare Figure 9.6 with the plots of the ICs from Chapter 8 (p. 121). This difference is due both to the masking effect in finite samples and also to to the iterative nature of the fitting process. It is apparent that the SC is a more accurate predictor of the influence of a point than the empirical IC.

9.5.2 Extreme sensitivity

This chapter has demonstrated that there are a large number of conditions (starting weights, minimization routine, stopping criteria, encoding of target values, numerical precision), such that altering any one of them will very likely lead to a different solution. The converse of the dependence on the starting weights is also easy to show; if the starting weights are held constant and the data moved it is easy to see that the MLP is not translation or rotation or scale invariant.

In addition, it has been demonstrated that an MLP model is extremely sensitive to outlying points in the data. This is the case whether the MLP is over–parameterized or is only minimally parameterized to perform the required classification.

In calculating the mean, say, a single outlying observation, x^*, may move $\bar{x}_n$ without bound. However, if x^* is fixed and n is increased then the influence of x^* will be diminished (assuming that there are no other outliers). This is not the case with an MLP. For a data set such as Figure 9.5 plot 6, increasing the number of observations from the rectangular regions (even by a factor of 10) does not diminish the impact of the single outlying point.

CHAPTER 10

A ROBUST FITTING PROCEDURE FOR MLP MODELS

10.1 INTRODUCTION

We have considered the influence and sensitivity curves for the MLP classifier; we now consider how the shapes of these curves can lead to strategies for robust classification. We now show that an MLP with a ρ_c penalty function is more resistant to outlying observations by the addition of a hidden layer[1]. That is, the $P.1$ model (logistic regression) and the $P.1.1$ model behave quite differently. In addition the $P.1.1$ model is more robust than several robust logistic regression models in the literature. Next, modifications of the MLP model are considered to increase its resistance and a new and highly effective method of robustifying the MLP is introduced. The procedure is contrasted with that of Hunt et al. (1992) and it is argued that the correct space in which to site the robust modification is that of the linear predictors or inputs to the first hidden layer, $y = \omega^\tau x$, not the feature space. In addition the question of model diagnostics is considered and a diagnostic tool is introduced.

Multi–layer perceptron models can readily be shown to give highly variable results with as little as a single outlying or misclassified point. Such models are said to lack the properties of "robustness" or "resistance" (Section 8.2, p. 122). For

---

[1] The differing shapes of the influence curves for MLP models with and without a hidden layer were discussed in Section 8.6, p. 136. It was suggested there that the shapes of the influence curves would lead to a $P.1.1$ MLP being more robust than a $P.1$ MLP, and we now establish that this is the case.

example, Figure 9.5 (p. 148) shows an MLP model, with one hidden–layer unit, trained to separate 2 classes in a 2 dimensional feature space. The movement of a single point in the "+" class has a marked effect on the position of the decision hyperplane. Figure 9.7 (p. 150) shows the same configuration and an MLP model with two hidden–layer units. In this case the effect on the MLP is even more drastic, with one of the hidden layer units devoted to chasing the aberrant point.

Robust fitting procedures for regression MLP models have been considered by Movellan (1990). It is suggested that the $t_n - \text{mlp}(x_n)$ term in

$$\psi_{\omega_r}(x) = \frac{\partial}{\partial \omega_r} \rho_l(x, \omega)$$

$$= \sum_{n=1}^{N} \{t_n - \text{mlp}(x_n)\} \frac{\partial}{\partial \omega_r} \text{mlp}(x_n, \omega)$$

(equation (8.3) on p. 128) be replaced by $\psi(t_n - \text{mlp}(x_n))$ where ψ is Tukey's biweight function (Section 8.4, p. 124). Similarly Bishop (1995a) suggests using a Minkowski penalty function

$$\rho(x, \omega) = \sum_{n=1}^{N} |t_n - \text{mlp}(x_n)|^r,$$

where $r < 2$, instead of the sum of squares error term ρ_l.

Kärkkäinen and Heikkola (2004) introduce more general fitting methods from non-smooth optimization and investigate MLP models with penalty terms of the form $||z^* - t||_\alpha^\beta$. They investigate the behavior of the MLP for various values of $\{\alpha, \beta\}$ and conclude that $\alpha = \beta = 1$ (L^1 regression) leads to a more robust solution.

Chen and Jain (1994) consider a penalty term based on Hampel's redescending tanh estimator. The cut-off values $\{a, b\}$ are made step dependent and are recalculated at each step using a function of the order statistics of the absolute residuals. See also Liano (1996) and Pernía-Espinoza et al. (2005) for robust modifications to the penalty term.

These suggestions are directed towards limiting the influence of the residual in a regression problem. They may not be relevant to a classification problem where the ρ_c penalty function may be more appropriate than the ρ_l function and where the influence of the position in feature space may raise more problems than the influence of the residual which will be bounded for most classification models. Robust MLP classification has not been considered in the literature with the exception of Hunt et al. (1992). The proposal outlined by Hunt et al. is considered in Section 10.4.1, p. 169.

10.2   THE EFFECT OF A HIDDEN LAYER

As shown in Section 4.6 (p. 40), there are a number of connections between the MLP model and other statistical models. The MLP model of size $P.1$ with ρ given by a Bernoulli negative log likelihood (equation (4.6), p. 40), f_Q chosen as a logistic activation function and $t \in \{0, 1\}$ is identical to the logistic regression model (Cox, 1958; Day and Kerridge, 1967). Also, with some parameter redundancy, this is the same model as $P.2$ with ρ given by ρ_c, f_Q chosen as the softmax function and t taking the values $(0, 1)$ and $(1, 0)$.

In addition, in Section 8.6 (p. 136) it was shown that the addition of a hidden layer to an MLP with a ρ_c penalty function and a softmax activation function has a major change on the shape of the ICs. For such an MLP of size $P.2$ (no hidden layer) the influence curves are unbounded in every direction (Figure 8.8, p. 140) whereas for the $P.1.2$ MLP they are either bounded or redescending in all directions except parallel to the decision boundary (Figure 8.4, p. 133, and Figure 8.5, p. 134). This suggests that a $P.1.1$ MLP will be more robust to outlying points than the logistic model MLP $P.1$.

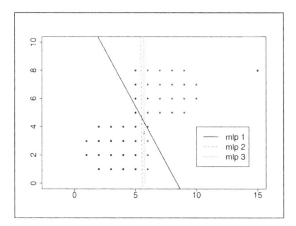

Figure 10.1 A data set of 101 examples (some at the same coordinates) in two classes, labeled "0" and "1." There is (perhaps) one outlying point in class "0." The decision boundaries for MLP models 1, 2 and 3 are plotted.

To illustrate this we consider a small numerical experiment[2]. The data set consists of 101 points in two classes and is shown in Figure 10.1. We consider three MLP models:

1. MLP 1 is of size 2.1 and has a ρ_c penalty function;

2. MLP 2 is of size 2.1 with a ρ_l penalty function;

3. MLP 3 is of size 2.1.1 with a ρ_c penalty function.

Examining Table 10.1 we see that there is a distinct difference in the error rates between the penalty functions for a single layer MLP with a ρ_l penalty function versus one with a ρ_c penalty term (MLP 1 versus MLP 2), illustrating the argument of Section 8.6, p. 136. When we consider MLP 3, we see that the addition of a hidden layer has made the MLP less sensitive to the outlier than MLP 1. We suspect that this robust behavior, along with "early stopping" caused by long training times, has been central to the success of MLP models in many practical examples. Table 10.1 gives the results for several classification methods on this data set.

In this chapter we consider ways to limit the influence of points that are separated from the main body of the data. However, another way of seeing this data set is to

---

[2]This example was posted to `comp.ai.neural-nets` by W. Sarle. MLP 1 and 2 recapitulate a part of Sarle's example.

view it as having 7 outliers, not 1. In this case, MLP 1 would appear to be finding the correct decision boundary and MLP 2 and 3 are overfitting the data. Clearly, the identification of outliers depends on what we assume about the data set, and in particular, for structured data, such as in Figure 10.1, outliers may not be points that are far from the main body of the data.

Table 10.1   The number of misclassified observations, from the data set illustrated in Figure 10.1, for several classification methods.

method	misclassification matrix	
linear discriminant analysis	$\begin{pmatrix} 46 & 4 \\ 4 & 47 \end{pmatrix}$	
linear regression	$\begin{pmatrix} 46 & 4 \\ 4 & 47 \end{pmatrix}$	
ρ_c with logistic activation function and no hidden layer (logistic regression)	$\begin{pmatrix} 47 & 3 \\ 4 & 47 \end{pmatrix}$	MLP 1
ρ_l with logistic activation function and no hidden layer	$\begin{pmatrix} 50 & 0 \\ 1 & 50 \end{pmatrix}$	MLP 2
ρ_c with logistic activation function and a hidden layer	$\begin{pmatrix} 50 & 0 \\ 1 & 50 \end{pmatrix}$	MLP 3
ρ_l with logistic activation function and a hidden layer	$\begin{pmatrix} 50 & 0 \\ 1 & 50 \end{pmatrix}$	

10.3   COMPARISON OF MLP WITH ROBUST LOGISTIC REGRESSION

The question of robust logistic regression has received some attention in the literature. We briefly consider the model introduced by Cox and Pearce (1997), a "Huber" model, and the model introduced by Pregibon (1982).

Cox and Pearce (1997) modify the usual logistic regression model,

$$y = \frac{\exp(\omega^T x)}{1 + \exp(\omega^T x)}, \tag{10.1}$$

to

$$y = \frac{k\{c_1 + c_2 \exp(\omega^T x)\}}{1 + \exp(\omega^T x) + k\{c_1 + c_2 \exp(\omega^T x)\}} \tag{10.2}$$

where k, c_1, and c_2 are constants assigned by the user, and the k parameter is the ratio of the priors.

The influence curves (ICs) for the logistic model are described in Section 8.6 (see particularly Figure 8.8, p. 140). In that section it was shown that the ICs for the logistic model are unbounded with two exceptions: the IC for ω_0; and the IC for ω_p along a path perpendicular to the x_p axis.

Following the argument of Section 8.5 (p. 128) we can calculate the influence curves for (10.2) by evaluating the derivative for a point κ as κ moves along paths either parallel or perpendicular to the decision boundary. It can be seen that the ICs will be unbounded along the paths parallel to $\omega^T x = 0$ but will redescend to 0 along paths perpendicular to $\omega^T x = 0$. Thus (10.2) should be more robust than the standard logistic model.

Another approach to robust logistic regression is to fit (10.1) but to weight each observation by

$$w_n = \frac{\psi[\rho_c\{t_n, z^*(x_n)\}]}{\rho_c\{t_n, z^*(x_n)\}},\tag{10.3}$$

where ψ is chosen to limit the influence of atypical observations. We could choose ψ as

$$\psi(x) = \begin{cases} -k & \text{when } x < -k \\ x & \text{when } |x| \le k \\ k & \text{when } x > k, \end{cases}$$

giving a "Huber" estimator[3]. This estimator provides resistance to observations that are poorly fitted by the model (it limits the *influence of the residual*) but does not provide resistance to outlying observations in feature space that may exert a large influence on the fitted model (it does not limit the *influence of the position in feature space*).

Pregibon (1982) also considers this question and suggests fitting the model by minimizing

$$\sum_n v(x_n)\psi\left\{\frac{\mathbb{D}(w^T x_n, t_n)}{v(x_n)}\right\},$$

where $\mathbb{D}$ is the deviance (see Section 5.2, p. 58) and v is chosen to down-weight observations that have an undue weight on the fit. v is a function of the diagonal entries of the projection or influence matrix considered in Section 10.5 (p. 172). Cox and Ferry (1991) compare Pregibon's resistant fitting with an earlier version of their robust model (on a simulated example which we revisit in the following section) and conclude that they are equally resistant to outliers.

10.3.1   A simulation study

We have implemented the following five models:

1. logistic regression (10.1);

2. an MLP of size $P.1$ with a ρ_l activation function (linear regression);

3. the robust model of Cox and Pearce (1997) (10.2);

4. an MLP of size $P.1.1$ with a ρ_c activation function; and

5. the Huber robust logistic regression model (10.3).

---

[3]This model is provided by the *S-PLUS* function `robust(binomial)`, which chooses $k = 1.345$ as a default.

Logistic regression (10.1) is a generalized linear model and software to fit the model is widely available. We have used both the `glm` function in *S–PLUS* and fitted an MLP model of size *P*.1 to get essentially the same answers[4]. We fitted the *P*.1 and *P*.1.1 MLP models using a Fortran subroutine linked to *S–PLUS*. We implemented Cox and Pearce's model (10.2) in *S–PLUS*. Model (10.3) is available as an option to the `glm` command in *S–PLUS*.

To test the models the simulation study of Cox and Pearce (1997) was replicated. A training data set of 100 bivariate Gaussian observations (x_1, x_2), was generated with $\mu = (0, 0)$ and

$$\Sigma = \begin{pmatrix} 1 & \sigma_{1,2} \\ \sigma_{1,2} & 1 \end{pmatrix},$$

with the covariance $\sigma_{1,2}$, taking one of a number of fixed values. The data were then randomly allocated to one of two groups (0 and 1) via a Bernoulli distribution with

$$P(t = 0) = \exp(x_1 + x_2)/(1 + \exp(x_1 + x_2)).$$

An additional 10 points in class 0 were situated at a number of positions to test robustness against outliers. The models were evaluated in terms of the misclassification rates and no attempt has been made to evaluate the accuracy of the estimates of the posterior probabilities of class membership.

Figure 10.2 gives the decision boundaries and resubstitution error rates (P_e^A, Section 5.2, p. 54) for each of the four models for the case $\sigma_{1,2} = 0.7$ and outlying points at $(-4, -4)$. We found that Cox and Pearce's model depended for its robustness on a low tolerance value (the value at which the minimization iterations are terminated) and thus on a version of early stopping (see Section 5.4, p. 62). Cox and Pearce give a set of GLIM macros (Francis et al., 1993) to fit the model and by varying the tolerance in *S–PLUS* we can replicate the output of a GLIM routine with a tolerance of 0.001. However, with a tolerance of 0.0001 it appears that the model is subject to overtraining.

Reporting the resubstitution error rates is a poor indication of the Bayesian error rate. The true error rate could be calculated by

$$P_e^{\text{bayes}} = P(g_1|g_0) + P(g_0|g_1),$$

where

$$P(g_1|g_0) = \int\!\!\int_{\omega^\tau x < 0} \frac{\exp(\omega^\tau x)}{1 + \exp(\omega^\tau x)} \phi(x) dx,$$

where ϕ is a bivariate Gaussian density with parameters μ and Σ. Unfortunately the integral involves the form $\int \exp(x^3)/(1 + \exp(1 - x^2)) dx$ which has no closed form solution. To avoid doing a numerical integral we report the error rate on an independent sample of 1000 observations from the same distribution. This gave an estimated error rate of 25.7 for the MLP model and 39.4 for Cox and Pearce's model (tolerance=0.001).

This is quite an intriguing result. Compared to two robust modifications of logistic regression, an MLP of size 2.1.1 does extremely well!

---

[4]Note that while these two models are the same, it is possible to get somewhat different fits as the fitting algorithm and initial parameter values are different in the standard implementations.

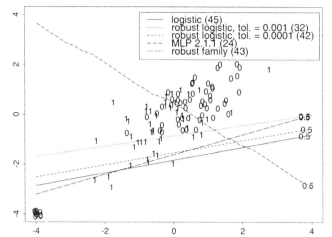

Figure 10.2   The two classes are represented by "0" and "1," respectively, and the group of 10 outliers at the point $(-4, -4)$ are shown at the bottom left of the figure. The data was generated using $\sigma_{1,2} = 0.7$. The positions of the outliers have been randomly perturbed for greater visibility. The decision boundaries for several of the models are shown. The number of errors for each model (not counting the ten additional points) is given in the legend.

For the simulation study the trials were repeated 10 times and the average error rates are reported in Tables 10.2 to 10.6. The optimal misclassification rates, using the "true" decision boundary of $x_1 + x_2 = 0$, are 35.18, 29.41, 26.98, 24.81, and 22.86 for the five values of $\sigma_{1,2}$ considered. These rates are for the main body of the data, not including the aberrant points. These problems range from the quite difficult, on which both the robust and standard models fail, to the somewhat easier, on which both methods do reasonably well. For example, the data generated with $\sigma_{1,2} = -0.7$ and outlying points at $(-4, -4)$ has an error rate of 41% on the full data set with the true decision boundary, while classifying all the observations to class 1 only increases the error rate to 45%.

A salient feature of Tables 10.3 to 10.6 is, moving from the bottom right of the table, the point at which the performance of the algorithm starts to exceed that of the logistic model (Table 10.2). We note several points arising from this study:

- while Cox and Pearce's model has ICs that redescend (in the same manner as the $P.1.1$ MLP model) it does not appear to be as robust. We have noted the dependence of the result on the tolerance level and in addition we note that two implementations of this model:

  - a *GLIM* implementation (Francis et al., 1993);
  - an *S–PLUS* implementation based on a a modified logistic model with a tolerance of 0.001;

  both give essentially the same answers[5]

---

[5]These results appear to be less robust than the results given by Cox and Pearce (1997) on the same problem. Currently we are at a loss to explain this anomaly.

- the MLP model with a hidden layer performs very well compared to the other models and could be used when a robust logistic regression is required. However in the harder problems ($\sigma_{1,2} < 0$) it fails to achieve an optimal error rate;

- the poor performance of the Huber estimator is due to the large number (10) of atypical points. With, for example, two or three atypical points at $(-4, -4)$ the Huber estimator gives a more satisfactory performance;

Table 10.2 The error rates for the logistic model (MLP of size $P.1$)

aberrant values at	$\sigma_{1,2}=-0.7$	$\sigma_{1,2}=-0.3$	$\sigma_{1,2}=0$	$\sigma_{1,2}=0.3$	$\sigma_{1,2}=0.7$
$-4,-4$	55.11	55.54	51.31	48.39	44.04
$-4,-3.5$	54.98	55.00	50.24	46.84	40.29
$-3.5,-4$	54.94	54.61	51.23	46.73	43.33
$-3,-3$	54.75	52.35	46.84	41.45	37.54
$-3,-1.5$	53.11	45.18	38.26	34.71	30.78
$-3,1.5$	42.36	33.24	30.44	27.77	25.07
$-1.5,-1.5$	51.09	39.52	34.96	31.03	28.69
$1.5,1.5$	36.14	30.10	27.46	25.33	23.92
$3,3$	36.37	30.16	27.60	25.38	23.99

Table 10.3 The error rates for the MLP of size $P.1$ with a ρ_l penalty function.

aberrant values at	$\sigma_{1,2}=-0.7$	$\sigma_{1,2}=-0.3$	$\sigma_{1,2}=0$	$\sigma_{1,2}=0.3$	$\sigma_{1,2}=0.7$
$-4,-4$	54.07	54.73	49.40	37.43	31.21
$-4,-3.5$	53.87	54.41	48.36	36.50	28.70
$-3.5,-4$	54.20	54.20	49.52	36.89	31.48
$-3,-3$	53.89	52.18	45.23	33.84	27.74
$-3,-1.5$	52.61	43.59	36.72	32.87	28.83
$-3,1.5$	42.39	33.61	30.55	27.70	24.73
$-1.5,-1.5$	51.08	39.03	34.00	29.16	27.45
$1.5,1.5$	36.35	30.32	27.72	25.54	24.21
$3,3$	36.54	30.39	27.75	25.57	24.17

10.4 A ROBUST MLP MODEL

While the MLP model has performed well by comparison with two robust logistic models, there is scope for improvement. We consider a weighting procedure designed to down-weight observations that lie at some distance from the main body of the data.

Table 10.4   The error rates for the robust model of Cox and Pearce (1997). The tolerance level is 0.001

aberrant values at	$\sigma_{1,2}=-0.7$	$\sigma_{1,2}=-0.3$	$\sigma_{1,2}=0$	$\sigma_{1,2}=0.3$	$\sigma_{1,2}=0.7$
$-4,-4$	55.09	55.50	51.19	48.38	43.00
$-4,-3.5$	54.93	55.63	50.59	44.03	38.00
$-3.5,-4$	55.24	54.50	50.27	46.04	42.65
$-3,-3$	54.94	52.85	46.49	39.56	35.51
$-3,-1.5$	53.13	44.66	37.65	34.02	30.42
$-3,1.5$	41.99	33.00	30.32	27.79	24.87
$-1.5,-1.5$	51.03	38.84	33.91	29.92	27.81
$1.5,1.5$	36.22	30.25	27.57	25.33	23.93
$3,3$	36.31	30.20	27.55	25.51	23.97

Table 10.5   The error rates for the MLP of size $P.1.1$ with a ρ_c penalty function.

aberrant values at	$\sigma_{1,2}=-0.7$	$\sigma_{1,2}=-0.3$	$\sigma_{1,2}=0$	$\sigma_{1,2}=0.3$	$\sigma_{1,2}=0.7$
$-4,-4$	54.07	54.73	49.41	37.43	28.98
$-4,-3.5$	53.87	54.41	48.36	36.50	28.70
$-3.5,-4$	54.21	54.20	49.52	36.89	29.02
$-3,-3$	53.89	52.18	45.23	33.17	27.74
$-3,-1.5$	52.61	43.59	36.72	32.87	28.67
$-3,1.5$	42.39	33.60	30.55	27.70	24.73
$-1.5,-1.5$	51.08	39.03	33.99	29.17	27.45
$1.5,1.5$	36.35	30.34	27.63	25.54	24.21
$3,3$	36.54	30.39	27.75	25.58	24.70

Table 10.6   The error rates for the Huber robust logistic model.

aberrant values at	$\sigma_{1,2}=-0.7$	$\sigma_{1,2}=-0.3$	$\sigma_{1,2}=0$	$\sigma_{1,2}=0.3$	$\sigma_{1,2}=0.7$
$-4,-4$	55.15	55.54	51.31	48.39	44.04
$-4,-3.5$	54.98	55.00	50.23	46.84	40.29
$-3.5,-4$	54.95	54.61	51.23	46.72	43.33
$-3,-3$	54.75	52.35	46.84	41.45	37.54
$-3,-1.5$	53.10	45.18	38.26	34.73	30.80
$-3,1.5$	42.37	33.25	30.38	27.74	24.98
$-1.5,-1.5$	51.09	39.52	34.97	31.00	28.69
$1.5,1.5$	36.14	30.08	27.46	25.35	24.01
$3,3$	36.37	30.24	27.56	25.41	23.99

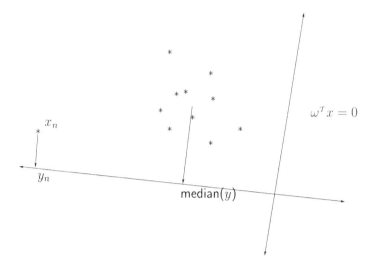

Figure 10.3   The line $\omega^T x = 0$ is the decision boundary for a hidden layer unit. The point x_n is an outlier from the cloud of points and y_n (its projection onto the normal to $\omega^T x = 0$) is far from the median of the ys and so the point x_n will have a low weight w_n.

The basic idea for implementing a robust fitting procedure is as follows: for each class and for each unit in the hidden layer, we calculate a weight function, based on robust estimates, to downgrade atypical observations (see Figure 10.3). To do this for a given hidden layer unit and the qth class we calculate $y_{n_q} = \omega^T x_{n_q}$ for $n_q = 1, \ldots, N_q$ the number of observations in the qth training class. We then calculate the class median, $\text{median}(y : y \in q^{th} \text{ class }) = \text{median}_q(y)$ say, and median absolute deviation (mad), where

$$\text{mad}(y : y \in q^{th} \text{ class }) = \text{median}(|y_{n_q} - \text{median}_q(y)|) = \text{mad}_q(y) \text{ say,}$$

with the residual r_{n_q} given by,

$$r_{n_q} = |y_{n_q} - \text{median}_q(y)|$$

and the scaled residual s_{n_q}, by

$$s_{n_q} = r_{n_q}/(\text{mad}_q(y)/0.6745). \tag{10.4}$$

The mad is divided by 0.6745 so that it achieves an asymptotic efficiency of 95% at the Gaussian distribution (Huber, 1981).

The weight function used here is

$$w_{n_q} = \begin{cases} 1 & s_{n_q} \leq 2 \\ 0 & s_{n_q} \geq a \\ 1 - \frac{1}{a-2}(s_{n_q} - 2) & 2 < s_{n_q} < a, \end{cases} \tag{10.5}$$

where a is a parameter that must be set. w_{n_q} is defined for an observation x_{n_q} in class q and for a particular unit in the hidden layer. However, the current implementation takes the weight, w^*, as the minimum of the weights over the units in the hidden layer and over successive iterations. Hence, once a direction is found in which an observation is atypical, the observation continues to be down–weighted. For each x_{n_q} in the training set, the weight w^* is added to the MLP model to give the following penalty function for a single observation:

$$\rho_n = -\sum_{q=1}^{Q} w^* t_q \log\left(\frac{z_q^*}{t_q}\right).$$

10.4.1   A "fuzzy set" approach

For the purpose of comparison we consider another weighting scheme for robustifying the MLP model. Hunt et al. (1992) propose improving the generalization ability of an MLP by including, in the objective function, information about how far an observation x is from the class means. To do this, they use the Mahalanobis distance

$$D_q^2(x) = (x - \hat{\mu}_q)^T \hat{\Sigma}_q^{-1}(x - \hat{\mu}_q),$$

where $\hat{\Sigma}_q$ and $\hat{\mu}_q$ are the sample covariance matrix and mean for the qth class. They then take a decreasing function of D_q^2 such as

$$w_q(x_n) = \frac{2}{1 + \exp\{D_q^2(x_n)\}},$$

which they refer to as a "fuzzy set membership function."

The appropriateness of the D^2 metric depends on the population distribution. If the distribution is not symmetric and unimodal with finite first and second moments, then $\hat{\mu}$ and $\hat{\Sigma}$ will not be sensible location and scale estimators. We note that Hunt only claims[6] that the method works well for ellipsoid and "banana" shaped classes.

Consider a two–class problem (C_1 and C_2) with $x \in C_1$. We follow Hunt's example[7] and use a ρ_l penalty function, which will be

$$\rho_l = \frac{1}{2}w_1(x)\{t_1 - z^*(x)\}^2 + \frac{1}{2}w_2(x)\{t_2 - z^*(x)\}^2.$$

Now suppose that x is closer to the mean of C_2 than C_1 so that, with Hunt's weight function, $w_1(x) < w_2(x)$. However, the contribution of x to the total error would be less if $w_2(x)$ was set to the same value as $w_1(x)$. In other words, nothing is gained by having a separate weight for each output unit. A better effect could be achieved by setting

$$w(x_n) = \frac{2}{1 + \exp\{D_c^2(x_n)\}},$$

where $D_c^2(x_n)$ is the Mahalanobis distance for the known class for x_n.

---

[6] This was communicated in a presentation at the Third Australian Conference on Neural Networks, 1992.

[7] There is no reason why Hunt's weighting scheme could not be applied to other penalty terms such as ρ_c.

10.4.2 Breakdown bounds and efficiency

It is desirable that an estimator have both a high breakdown point and high efficiency. The breakdown point, ϵ^*, is defined as the largest fraction of gross errors that cannot move the estimate beyond all bounds. $\epsilon^* = 0$ for $\hat{\mu}$ and $\hat{\Sigma}$, as one observation out of n can arbitrarily affect the estimate and $1/n \to 0$ as $n \to \infty$. For the median, however, $\epsilon^* = 1/2$ $(1/2 - 1/n \to 1/2$ as $n \to \infty)$ which is the maximum possible for any estimator.

The other quantity of interest is the efficiency or asymptotic variance. This is the ratio of the variance of the robust estimate to the variance of a standard estimate when the data are drawn from the same distribution, often the Gaussian. The efficiency depends on both the estimates of the residuals and the weight function.

10.4.3 Discussion of Hunt's procedure

Hunt et al.'s procedure is related to a form of robustification called classical outlier rejection (Rousseeuw and Leroy, 1987). This procedure works by calculating the D^2 value for every observation, and removing the observations that have a large D^2 value. This has the problem[8] that a single far–away outlier will inflate $\hat{\Sigma}$, giving other outliers a small D^2. Because of this "masking" effect, the breakdown point of classical outlier rejection is only $2/n$.

Hunt's procedure also has an unnecessarily low efficiency at the Gaussian due to the severity of its weight function. From Table 10.7, we can see that observations at only one standard deviation from the mean have been down–weighted by ≈ 0.5.

standard deviations	Hunt's method	robust w
0	1	1
1	.538	1
2	.2384	1
3	0.0948	0.667
4	0.036	0.333
5	0.013	0

Table 10.7   A comparison of the weights for Hunt's procedure and the proposed robust weighting (10.5) at various distances (measured in standard deviations) from the mean of a Gaussian population.

Due to these problems, we do not recommend that Hunt's procedure be used in its current form. However, it could be improved by using robust estimators for μ and Σ and using a less severe weight function. To test this we implemented a version of Hunt's procedure using the minimum volume ellipsoid covariance estimator. This gives a covariance matrix defined by the ellipsoid with minimum volume among those ellipsoids that contain $\lfloor \{(N + P + 1)/2\} \rfloor$ of the data points (Rousseeuw and van Zomeren, 1990). See Section 10.4.6 (p. 172), where we apply Hunt's procedure and our robust version of Hunt's procedure to the example from Section 10.3.1, and Tables 10.9 and 10.10 for the error rates. The changes made to the procedure appear to have made it more robust.

---

[8] As $\hat{\Sigma}$ is readily affected by a single observation (more so than $\hat{\mu}$), one outlier may greatly inflate the estimate.

10.4.4   Discussion of the robust procedure

The robust MLP has good breakdown properties as it calculates the median and the median absolute deviation for each class at each iteration.

The efficiency is somewhat more intractable. However, in other contexts the chosen combination of outlier detection and rejection procedures has performed well. Hampel (1985) studied the variances of estimates of the mean after the rejection of outliers using a number of test and rejection rules. The study considered a number of distributions including the Gaussian and the Cauchy and various degrees of contamination by outliers (0% to 50%). Model (10.4) performed well and (10.5), with $a = 3$, was chosen to be continuous and a trade–off between efficiency at the Gaussian and efficiency at heavier tailed distributions. We have followed the recommendations of this study.

The proposed robust procedure amounts to a reweighting of projections of the data. Diaconis and Freedman (1984) give results on the distributions of low–dimensional projections of high–dimensional distributions. A projection of a point in $\mathbb{R}^P$ onto $\mathbb{R}^1$ is a weighted sum of the coordinates and thus, with some restrictions on the distribution of the coordinates of a random variable, central limit type theorems can be applied[9]. These results have been important in projection pursuit regression and clustering via a characterization of the "most interesting" projection of a data cloud in $\mathbb{R}^P$ as the "least Gaussian" projection. Huber (1985) summarizes these results as

> ... for most high–dimensional data clouds most low–dimensional projections are approximately [Gaussian].

As the projections may be approximately Gaussian, making the weighting asymptotically efficient at the Gaussian will in general be desirable. Hence capitalizing on the similarities of MLP models and projection pursuit methods and robustifying the projections has the potential to be an effective way of dealing with high–dimensional data of unknown distribution.

10.4.5   Sensitivity curves

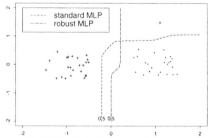

Figure 10.4   As in Figure 9.7, MLPs of size 2.2.1 were fitted to this data set. Two decision boundaries are shown, one for a standard MLP, which chases the aberrant point, and one for the robust MLP, which does not.

---

[9]The Levy–Lindberg central limit theorem will probably not be applicable as the coordinates are unlikely to be independent and identically distributed.

We consider the effect of the robust modification of the MLP on the shape of the sensitivity curves. In Figure 10.4, we consider the same scenario as in Figure 9.7 (p. 150) except that the MLP is now fitted robustly. Clearly the single point now has little effect on the final model when the robust procedure is adopted.

The SC (Section 9.2, p. 144) for the parameter ω_2 has been plotted in Figure 9.6 (p. 149) for the standard MLP. In Figure 10.5 a contour plot is given for the SC for ω_2 for the robustly fitted MLP. Comparing Figures 9.6 and 10.5 it can be seen that the influence of a single point in some regions of feature space will have a much larger effect for the standard MLP than for the robustly fitted MLP. Note that, unlike the standard MLP, the SC does not follow the lines denoting the limits of linear separability. As the fitted MLP model is highly dependent on the starting values, the illustrated SC only reflects the MLP model with a particular set of starting values. It would be possible to remove the effect of the starting values and get a fuller picture of the shape of the SC by averaging over a number of fits from different starting values.

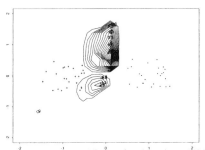

Figure 10.5 The sensitivity curve for weight w_2 for the robust MLP of size 2.2.1.

10.4.6 The example revisited

We reconsider the example of Section 10.3.1 (p. 163), but now using the robust MLP, Hunt's procedure, and our robust version of Hunt's procedure. In order to make the problem a little harder we now add a single additional point at $(-20, -20)$ in addition to the 10 outlying points. Table 10.8 shows the results of using the MLP model. Table 10.9 shows the results of an MLP of size $P.1.1$ with Hunt's procedure while Table 10.10 shows Hunt's procedure with the robust modifications.

Both the robust version of Hunt's procedure and the robust MLP model perform well in this setting. Figure 10.6 shows the decision boundary for these two models in the case of $\sigma_{1,2} = -0.7$ and aberrant points at $(-4, -4)$. Hunt's procedure failed in this case and the decision boundary does not intersect the plot. The reason for this is the additional point at $(-20, -20)$ (not shown in the figure) which greatly inflates the non-robust estimate of Σ.

10.5 DIAGNOSTICS

For linear models a number of diagnostic procedures are available. It is fairly standard to have access to diagnostic output such as the following:

Table 10.8   The error rates for the robust MLP model of size $P.1.1$.

aberrant values at	$\sigma_{1,2}=-0.7$	$\sigma_{1,2}=-0.3$	$\sigma_{1,2}=0$	$\sigma_{1,2}=0.3$	$\sigma_{1,2}=0.7$
-4,-4	37.75	32.11	28.19	25.98	24.00
-4,-3.5	37.72	32.11	28.19	25.98	24.00
-3.5,-4	37.75	32.11	28.19	25.98	24.00
-3,-3	37.75	32.11	28.19	25.99	24.05
-3,-1.5	37.74	32.50	28.19	26.24	24.57
-3,1.5	42.40	35.23	31.00	27.61	24.98
-1.5,-1.5	38.88	34.15	31.86	29.70	25.57
1.5,1.5	38.00	31.38	28.06	25.78	23.92
3,3	37.75	32.15	28.08	25.88	23.96

Table 10.9   The error rates for Hunt's procedure, MLP $P.1.1$, ρ_l. The data set has an additional point at (-20,-20).

aberrant values at	$\sigma_{1,2}=-0.7$	$\sigma_{1,2}=-0.3$	$\sigma_{1,2}=0$	$\sigma_{1,2}=0.3$	$\sigma_{1,2}=0.7$
$-4,-4$	50.36	48.03	39.55	28.63	27.25
$-4,-3.5$	50.55	47.03	36.91	28.67	27.22
$-3.5,-4$	50.43	47.73	39.00	28.56	27.23
$-3,-3$	50.38	48.10	41.79	28.70	27.59
$-3,-1.5$	49.16	41.95	34.86	30.93	28.17
$-3,1.5$	41.28	35.15	30.86	28.97	28.62
$-1.5,-1.5$	50.63	47.60	42.02	37.44	28.86
1.5,1.5	39.65	35.79	31.37	29.31	26.71
3,3	39.62	35.77	31.21	29.27	26.72

Table 10.10   The error rates for the robustified Hunt's procedure. The data set has an additional point at (-20,-20).

aberrant values at	$\sigma_{1,2}=-0.7$	$\sigma_{1,2}=-0.3$	$\sigma_{1,2}=0$	$\sigma_{1,2}=0.3$	$\sigma_{1,2}=0.7$
$-4,-4$	37.33	32.52	29.83	28.04	25.46
$-4,-3.5$	37.36	32.02	29.56	26.94	25.49
$-3.5,-4$	38.49	31.93	29.60	26.84	25.75
$-3,-3$	37.09	32.01	29.73	27.02	25.38
$-3,-1.5$	37.13	31.88	30.34	31.06	26.83
$-3,1.5$	40.48	32.30	30.39	26.69	25.47
$-1.5,-1.5$	36.97	32.57	32.96	31.87	27.68
1.5,1.5	38.24	32.83	31.16	28.86	26.70
3,3	37.16	31.86	29.90	28.68	25.89

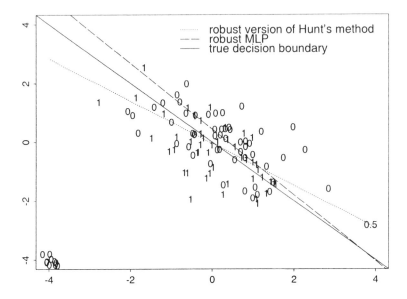

Figure 10.6   The two classes are represented by "0" and "1," respectively, and the group of 10 outliers at the point $(-4, -4)$ are shown at the bottom left of the figure. The data was generated using $\sigma_{1,2} = -0.7$. An additional outlier at $(-20, -20)$ is not shown. The positions of the outliers have been randomly perturbed for greater visibility. The decision boundaries for several of the models are shown.

- an N by P matrix of leave-one-out parameters, that is a matrix with entries $\omega_{(n)p}$, ω_p calculated with the nth observation omitted;

- a vector of the N studentized residuals (also called the *jackknifed* residuals)

$$\frac{t_n - z^*(x_n)}{\sqrt{\text{var}(t_n - z^*(x_n))}},$$

where $z^*(x_n)$ is z^* calculated without the point x_n, and then evaluated at x_n;

- the N diagonal values of the projection matrix for the fitted model.

The diagonal elements of the projection matrix $P_X = X(X^T X)^{-1} X^T$ have a direct interpretation as $\partial z^*/\partial t$, the rate of change of the fitted values for a change in t (t would be a continuous variable in this case). In the case of a linear model the leave-one-out estimates can be calculated without having to refit the model.

For generalized linear models a common practice (Hastie and Pregibon, 1992) is to take one step from the fitted model. That is, one observation is omitted and one step of the iteratively re-weighted least squares algorithm (Section A.4, p. 249) is taken from the fitted model. This gives computationally cheap but effective estimates of both the leave-one-out parameter and the jackknifed residuals[10]. These outputs can be examined directly or they can be used to calculate Cook's distance diagnostic (Cook and Weisberg, 1982) showing the influence of deleting an observation in the parameter vector in an appropriate metric. As these are now on the same scale they can be plotted and observations with a large Cook's distance can be investigated further. This procedure can be carried out for logistic regression, the robust logistic model and also for Cox and Pearce's model, which is a generalized linear model.

However the simulated example shows a problem with these measures. As there are 10 outliers at the one coordinate position, omitting one of these points has a negligible effect on the parameter estimates. Consequently the Cook's distance for the outlying points is quite small. In order to detect the 10 points as outliers, subsets of points (with up to 10 points in the subset) would have to be deleted at a time. In the general case an all–subsets deletion algorithm would be required. Deleting groups of outliers is computationally intensive for linear and generalized linear models and the problem is compounded for iteratively fitted non-linear models such as the MLP.

The leave-one-out parameter estimates and residuals could be calculated but this does not seem at all practical in the case of the MLP as the model has to be refitted for each omitted observation. We can linearize the model at the fitted values and calculate the $N \times P$ matrix D where

$$D_{nr} = \frac{\partial \rho(x_n)}{\partial \omega_r},$$

which is a linearized estimate of P_X. However, consider the example shown in Figure 10.7. At the fitted model, the derivative for the outlying point (50) will be relatively small. However the point has clearly been quite influential in determining the final fit.

Hence tracking the derivatives through time may provide a better diagnostic tool than the matrix D. This will give a series of N curves for each parameter which can be plotted on the same axis. It is possible that highly influential points may be seen to increase in $\partial \rho(x_n)/\partial \omega_r$ beyond the value for the major body of the data. Figure 10.8 shows a plot for the model from Figure 10.7. It can be clearly seen that during the fitting process the point labeled 50 has had a great influence on the parameter ω_2. This is a point that should be investigated further as a potential outlier.

10.6   CONCLUSION

The MLP model $P.1.1$ has been demonstrated as being as robust as several robust logistic models in the literature.

In addition, two proposals have been made

1. a robust modification of Hunt et al.'s procedure; and

2. a more robust modification of the MLP model.

---

[10]In the case of S–$PLUS$ it allows the use of the same algorithm and code as in the linear model case.

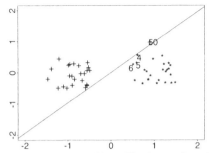

Figure 10.7   A two class classification problem. The decision boundary was formed by a 2.2.1 MLP. Note that several points, $\{4, 5, 6, 50\}$, have been labeled according to their row number in the data matrix.

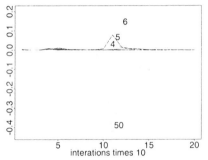

Figure 10.8   The derivatives for the parameter ω_2 during the course of fitting the model shown in Figure 10.7. The derivatives for points $\{4, 5, 6, 50\}$ are numbered.

The proposed robust modification is simple to implement, and appears to prevent overfitting of the decision boundary to identify atypical examples. It is applicable to MLPs with an arbitrary number of hidden layers and skip layers. However, as it involves two sorting steps, it is computationally expensive.

A simple and direct diagnostic tool has been suggested that will allow some interrogation of the fitted model to find highly influential points.

10.7   COMPLEMENTS AND EXERCISES

10.1   Redo the simple synthetic problem considered in Section 9.3.3 (p. 149) and elsewhere. Try a number of different starting values.

You will find that the robust MLP model will misclassify some of the points on one or the other of the two "wings" of the data set. This is a typical problem with structured data. A robust method may consider part of the structure to consist of outliers.

10.2   The robust modification of Hunt's method is most easily implemented by calculating a weight vector for each observation. This can be done as follows:

```
P<-dim(data)[[2]]
aa<-split(data,target[,2])
```

```
dim(aa[[1]])<- c(length(aa[[1]])/P,P)
aa1.cov <- cov.mve(aa[[1]])
aa1.mah <- mahalanobis(aa[[1]], aa1.cov$center, aa1.cov$cov)
dim(aa[[2]])<- c(length(aa[[2]])/P,P)
aa2.cov <- cov.mve(aa[[2]])
aa2.mah <- mahalanobis(aa[[2]], aa2.cov$center, aa2.cov$cov)
weights<-(2/(1+exp(c(aa1.mah,aa2.mah))))
```

You will have to use the `nnet` function from the `MASS` library as the current version of the `mlp` code does not allow for a weight vector.

If you try it on the data from the previous problem, you will see the same problem occurring. Using the `ellipse` function we can explore the reasons for this graphically (see Figure 10.9).

```
library(car)
ellipse(aa1.cov$cente, aa1.cov$cov, sqrt(qchisq(.5,2)))
ellipse(aa2.cov$cente, aa2.cov$cov, sqrt(qchisq(.5,2)))
```

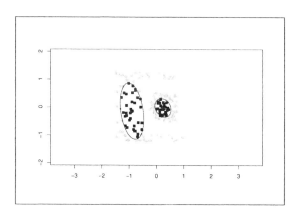

Figure 10.9   The centroids and ellipses for the data set of Exercise 10.2. In addition the points have been coded according to their weights. For further details see the script files.

10.3   The `MASS` library provides the Cushings data set (Aitchison and Dunsmore, 1975). Cushing's syndrome is a hypertensive disorder associated with over-secretion of cortisol by the adrenal gland. The observations are urinary excretion rates of two steroid metabolites and the type of syndrome:

- 'a' (adenoma);
- 'b' (bilateral hyperplasia);
- 'c' (carcinoma);
- 'u' for unknown.

The data set is 2–dimensional so it allows for easy visualization. Fit an MLP model and a robust MLP model to this data set and compare the results.

CHAPTER 11

SMOOTHED WEIGHTS

11.1 INTRODUCTION

We explore models and approaches that are applicable in the case where the individual observations are generated by sampling a smooth function at P points. In particular we consider the case where P is large, so that when the data are plotted they give the appearance of a continuous smooth curve[1]. This situation requires some modification of the usual analytic approach and has been considered by a number of authors. Ramsay and Silverman (1997) have coined the term "functional data analysis" for this type of analysis, while in the regression context is has been called "signal regression" (Marx and Eilers, 1999). Two modifications of the MLP are introduced, a basis function MLP (Section 11.2.1, p. 185) and a smoothness penalty term[2] (Section 11.2.2, p. 11.2.2). These modifications are discussed and their properties explored in a series of examples.

There are two approaches that are taken in the literature on signal or spectral observations. One is that of methods such as as penalized discriminant analysis (Hastie et al., 1995) and smoothed canonical variate analysis (Kiiveri, 1992), that add a roughness penalty term to the fitting procedure. The other is that of Furby et al. (1990) and Ramsay and Silverman (1997) who model the N observed curves

---

[1] All of the examples we consider in this chapter were generated by automated data collection processes so that the sampling of a large number of points was inexpensive.
[2] Some authors call this a "roughness penalty" as it penalizes roughness.

by N basis function expansions (B-spline, Fourier or some other appropriate basis) and then base their analysis on the coefficients of the basis function expansions. The selection of the number and position of the basis functions is clearly an important question in these methods. Furby et al. (1990) use a small number of adaptively selected B-spline basis functions while Ramsay and Silverman (1997, 1999) use a generous number of basis functions to model the data, and then apply a roughness penalty to the coefficients of the basis function expansions. As the number of basis functions is large, the basis function coefficients show serial correlation. Ramsay and Silverman thus substitute the continuous optimization problem of finding the penalty coefficient λ for the more difficult discrete optimization problem of determining the number of knots.

We consider the implications of treating the data as a function for modeling the variance. For the vector valued case we have $x(t) = \mu(t) + \epsilon$ where $\epsilon \sim N(0, \Sigma)$ and

$$\Sigma = Z\Lambda Z^{T}, \tag{11.1}$$

$$= \sum_i \lambda_i \zeta_i \zeta_i^{T}, \tag{11.2}$$

where Λ is a diagonal matrix containing the eigenvalues of Σ and the columns of Z contain the corresponding eigenvectors. We can then write $x(t) = \mu(t) + \sum_i \epsilon_i \zeta_i$, where $\mathbb{E}(\epsilon_i) = 0$, $\mathbb{E}(\epsilon_i^2) = \lambda_i$ and $\mathbb{E}(\epsilon_i \epsilon_j) = 0$ for $i \neq j$. Similarly, for the function valued case we have $x(t) = \mu(t) + \sum_i \epsilon_i \zeta_i(t)$ where the same conditions on ϵ_i hold as above, but now the $\zeta_i(t)$ are eigenfunctions of the covariance function

$$\Sigma(s, t) = \sum_i \lambda_i \zeta_i(s) \zeta_i(t), \tag{11.3}$$

and the $\zeta_i(t)$ are orthogonal in the L^2 sense (Rice and Silverman, 1991).

We see an example of this in Figure 11.1 which shows data from the HyMap example described in Section 11.3.3 (p. 193). A single observation is plotted and it appears reasonable to model the observations, in this case, as a mean curve plus a sum of curves rather than a mean curve plus uncorrelated Gaussian noise. The figure also shows the mean curve $\mu(t)$ plus and minus the first two eigenfunctions, calculated using the methodology of Ramsay and Silverman (1997). Note that, as is common in applications, the first eigenfunction appears to be largely due to the magnitude of the observations. The HyMap data is a "hyper-spectral" image, with multiple bands for every pixel. Figure 11.2 (p. 182) displays the first three eigenfunctions (or smoothed eigenvectors).

We briefly consider how estimates of these eigenfunctions were arrived at. We consider the functional case and write the inner-product as

$$\int \zeta_i(t) \zeta_j(t) dt = \; < \zeta_i, \zeta_j > .$$

The case where x is a vector is then notationally similar in terms of inner-products. The first principal component is found by

$$\underset{<\zeta,\zeta>=1}{\arg\max} < \zeta, \Sigma\zeta >$$

where Σ is the covariance of the mean centered data. This can be written as

$$\underset{\zeta}{\arg\max} \frac{< \zeta, \Sigma\zeta >}{< \zeta, \zeta >} \tag{11.4}$$

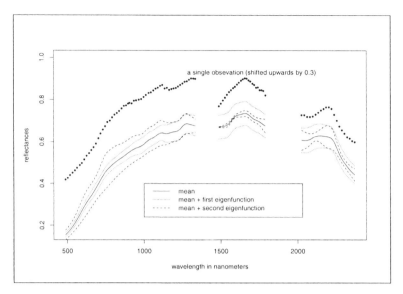

Figure 11.1   The data is the HyMap example (described in Section 11.3.3, p. 193). The plot shows the mean curve $\mu(t)$ plus and minus the first two eigen-functions of the covariance function. For comparison, a single observation (shifted upwards by 0.3) is also plotted.

as any scaling of ζ does not affect the value of (11.4). The jth principal component is then found by

$$\arg\max_{\zeta} \frac{<\zeta_j, \Sigma\zeta_j>}{<\zeta_j, \zeta_j>} \text{ such that } <\zeta_m, \zeta_j> = 0 \text{ for } m < j.$$

(11.4) leads to an integral equation

$$\int \Sigma(s,t)\zeta(t)dt = \lambda\zeta(s), \tag{11.5}$$

which becomes a matrix equation when the integral is solved by a numerical procedure (Ramsay and Silverman, 1997, §6) as will generally be the case.

We note that the model in which $x(t)$ is a smooth function plus additive noise is incorporated into this model by observing that the effect of white noise is to add a constant to each eigenvalue. The covariance function of white noise is the Dirac delta function $\delta(t-s)$ so (11.5) becomes

$$\int \{\Sigma(s,t) + \sigma^2\delta(t-s)\}\zeta(t)dt = \lambda\zeta(s)$$

$$\int \Sigma(s,t)\zeta(t)dt + \int \sigma^2\delta(t-s)\zeta(t)dt = \lambda\zeta(s)$$

which, by the properties of the delta function, gives

$$\int \Sigma(s,t)\zeta(t)dt + \sigma^2\zeta(s) = \lambda\zeta(s)$$

Figure 11.2   A functional principal component image of the HyMap data for Toolibin.

and so

$$\int \Sigma(s,t)\zeta(t)dt = (\lambda + \sigma^2)\zeta(s)$$

which is (11.5) with a constant value of σ^2 added to each eigenvalue.
Returning to the functional case we can penalize (11.4) as

$$\underset{\zeta}{\arg\max}\frac{<\zeta, \Sigma\zeta>}{<\zeta,\zeta> +\lambda||D^2\zeta||^2}, \qquad (11.6)$$

where D^2 is the second derivative operator. We require the boundary conditions that $D^2\zeta = D^3\zeta = 0$ at the ends of the interval[3]. With these conditions it can be shown, by a repeated integration by parts, that D^2 is self-adjoint[4] so that

$$||D^2\zeta||^2 = <D^2\zeta, D^2\zeta> = <\zeta, D^4\zeta>$$

[3]Different boundary conditions are applicable in the case where the function is assumed to be periodic. See Ramsay and Silverman (1997, §7.2).
[4]For a matrix, the adjoint is the conjugate transpose. So a self-adjoint matrix or Hermitian matrix is equal to its conjugate transpose $A = A^H = \overline{A^T}$ (Strang, 1988).

so (11.6) can be written as

$$\arg\max_{\zeta} \frac{<\zeta, \Sigma\zeta>}{<\zeta, (I + \lambda D^4)\zeta>}. \tag{11.7}$$

Apart from its use in determining the eigenfunctions (smooth eigenvectors) for models (11.2) and (11.3) (p. 180) an advantage of the formulation of (11.7) is that it can be readily recast for a smoothed version of discriminant analysis as

$$\arg\max_{\zeta} \frac{<\zeta, \Sigma_B\zeta>}{<\zeta, (\Sigma_W + \lambda D^4)\zeta>}. \tag{11.8}$$

To get a finite approximation to the D^2 penalty term we approximate the second derivative with the second forward difference operator matrix. Let D_1 be a matrix that applies the first difference operator to a vector,

$$D_{1_{ij}} = \begin{cases} 1 & i = j \\ -1 & i + 1 = j \\ 0 & \text{otherwise.} \end{cases}$$

Then D_2, the second difference matrix, is formed by

$$\underset{(P-2)\times P}{D_2} = \underset{(P-2)\times(P-1)}{D_1} \underset{(P-1)\times P}{D_1}$$

and we write

$$\underset{P\times P}{\Omega} = D_2^T D_2.$$

In the vector case D^T is the adjoint of D so that

$$||D_2\zeta||^2 = <D_2\zeta, D_2\zeta> = <\zeta, D_2{}^T D_2\zeta> = <\zeta, \Omega\zeta>.$$

We can then recast (11.8) as

$$\arg\max_{\zeta} \frac{<\zeta, \Sigma_B\zeta>}{<\zeta, (\Sigma_W + \lambda\Omega)\zeta>}. \tag{11.9}$$

See Kiiveri (1992) for a discussion of canonical variate analysis in high-dimensional spaces based on (11.9). In Section 11.3.3 (p. 193) a numerical implementation of (11.9) is discussed in connection with the HyMap data.

We can improve the interpretability of λ by replacing $\Sigma_W + \lambda\Omega$ by

$$\Sigma_W^\star = (1 - \lambda)\Sigma_W + \lambda\Omega\frac{\text{trace}(\Sigma_W)}{\text{trace}(\Omega)}.$$

This constrains λ to lie in $[0, 1]$ and ensures that $\text{trace}(\Sigma_W^\star) = \text{trace}(\Sigma_W)$ independently of λ so that the total variance of the data (as measured by the trace) is retained[5].

There is another related approach to functional CVA, namely the "penalized discriminant analysis" approach of Hastie et al. (1995). As noted in Section 3.3,

---

[5]This was suggested by Mark Berman (CSIRO) in a personal communication. It is used by Friedman (1989) in a different context.

p. 23, LDA on X and LDA on the fitted values from $\hat{T} = X\hat{\beta}$ give the same result. Similarly, (11.9) and a standard LDA on the fitted values from $T = X\beta + \lambda||D^2\beta||^2$ give the same result. This is suggestive of a considerable number of generalizations of LDA, for example, a penalized or smooth regression technique to be used as a building block for a penalized LDA. Many such techniques appear in the chemometrics literature or under the heading of "signal regression, " see, for example, Frank and Friedman (1993) or Ramsay and Silverman (1997)

11.2   MLP MODELS

Similar to the LDA models just discussed we have three notional MLP models:

1. the standard MLP model;

2. the functional MLP model where the data are smooth curves. In this model the sum $\omega^T x$ is replaced by the integral

$$\int \psi(t) X(t) dt;$$

3. a discrete approximation to the functional model – the data are modeled as smooth curves sampled at P points where P is relatively large.

In the case of the discrete approximation we unfortunately have no access to the curve X from which our datum is drawn. Our assumption that X is smooth is reflected in the fact that we model ψ as smooth. Thus the central change from the standard MLP model to the functional model is that ω_p weights fanning into a hidden layer unit are modeled as a function evaluated at the points p.

While it is possible to draw a C^∞ function through a given set of weights, we expect something more, we expect the weights to be visually smooth and not have the white noise appearance that may be associated with fitting P independent parameters.

Previously a number of penalty terms have been discussed (Section 5.4, p. 62), such as weight decay and weight elimination. We briefly consider what the effect of these penalty terms may be in the case of functional observations. Weight decay (Plaut et al., 1986) is the MLP equivalent of ridge regression. Its effect is to shrink the weights and thus smooth the boundaries of the decision region rather than to smooth $\psi(x)$. Weight elimination (Weigend et al., 1991) is a variable selection technique, and assumes an (improper) prior of a Gaussian contaminated with a uniform distribution. It essentially divides the weights into two groups, those from the uniform distribution that are not equal to 0, and those from the zero-centered Gaussian that can be considered to be equal to zero and removed. This is likely to result in ψ being a more spiky function rather than a smooth function.

Since the usual penalty terms are not appropriate, we consider a number of ways to ensure that the weight vector is smooth when the data are curves. Two novel methods that build the smoothness constraint into the MLP model or the fitting procedure are introduced in the following two sections but first we consider four direct ways of approaching the problem, namely:

1. *Do nothing.* As the vectors X are smooth functions, the resulting weights should be smooth and even if they are not, it may not matter. In prediction

(ie regression) problems it has been argued that modifications of the model to incorporate the serial correlation of the variables do not lead to a significant increase in accuracy; perhaps the same is true for classification problems;

2. *Fit a standard MLP* and then smooth the resulting weight vector;

3. *Smooth the weight vector* after each step;

4. *Impose a penalty term*, such as $(w_r - w_{r+1})^2$, that encourages the weight vector to be smooth.

Methods 1 and 2 will be considered in the examples and the conclusion. Neither of methods 3 or 4 is successful as the derivative term, $\partial \rho / \partial \omega$, is not modified (method 3) or only trivially modified (method 4). This means say, in the case of model 3, that the downhill step and the smoothing step may work in opposite directions. In practice it appears that this causes the minimization algorithm to be readily trapped in a see-saw motion.

11.2.1   A basis function expansion

In order to incorporate the smoothness constraint into the MLP, we model the unknown function ψ as

$$\psi \approx \sum_{k=1}^{K} \alpha_k B_k,$$

where $\{B_k\}_{k=1}^{K}$ is some set of basis functions. In this way, the number of estimated parameters will be reduced from $P + 1$ to $K + 1$ per hidden layer unit, where K is the number of basis functions used to model ψ (the bias term is not included in the basis function expansion).

Any set of convenient basis functions could be used including B-splines and truncated Gaussians. It appears preferable to use basis functions that have compact support so that the fitting procedure (iterative function minimization) only has to make local adjustments.

Note that even when using radially symmetric basis functions, this is not a probabilistic or radial basis function network[6] as we are not modeling the class densities. The classes are still separated with hyperplanes, as in the conventional MLP; we are simply modeling the decision boundary and so reducing the number of parameters needed to specify the hyperplanes.

To fit the model we generate a matrix B of size $K \times P$ such that

$$B[k, p] = \psi_k(p)$$

so B contains the K basis functions evaluated at the P points. The model is described by

$$z^* = f_Q(\Upsilon[1, \{f_H(A(1, Bx))^\tau\}^\tau]^\tau),$$

where A (of size $H \times K + 1$) is the matrix of weights to be estimated and the model can be fitted by any of the usual algorithms. In addition the model may be fitted

---

[6]Probabilistic networks as known under the name "kernel density estimation" in the large statistical literature on the area. This work has been largely duplicated by the neural network community. See Specht (1990a,b, 1991) for the neural network approach and Silverman (1986) for an earlier statistical approach.

with a penalty term, such as a weight decay term, on the α parameters to give a penalty term of the form

$$\rho + \sum_{\text{all } \alpha \text{ terms}} \alpha^2$$

The model inherits the properties of the chosen basis function. For example, if a spline basis is used then it is possible to have a great deal of control over the degree of continuity of the functions ψ. We can readily include discontinuities in the function (see the HyMap example) or in its derivatives by the choice of degree of the spline and the placement of the knots[7].

11.2.2   A smoothness penalty term

Another way to incorporate a serial smoothness constraint on the weights is to impose a penalty term of the form

$$\rho + \lambda \sum_h \int \left(\frac{\partial^2 \omega_h(x)}{\partial x^2} \right)^2 dx. \tag{11.10}$$

To get a finite approximation to this penalty term we approximate the second derivative with the second forward difference operator matrix D_2 and write

$$\Omega_{P \times P} = D_2^T D_2.$$

The appropriate penalty term is then $\rho + \lambda \omega^T \Omega \omega$, once again with the penalty not applied to the bias terms and there is a separate penalty term for the weights fanning into each hidden layer unit. In a $P.1$ MLP with a ρ_l penalty term it is equivalent to generalized ridge regression and to ridge regression if Ω is chosen as the identity matrix.

Bishop (1993) uses a penalty term of the form

$$\sum_n ||z_n^* - t_n||^2 + \lambda \sum_n \int \left(\frac{\partial^2 z_n^*}{\partial x^2} \right)^2 dx \tag{11.11}$$

for a regression network with linear output units. This can be seen to be a Tikhonov regularization term (see Section 5.7, p. 63). Note that different and unrelated quantities are being smoothed in (11.10) and (11.11). In (11.10) the differential operator is applied to the weight vector and in (11.11) it is applied to the outputs of the MLP. Now, it is possible for the output of the MLP to be smooth without the weights, considered as a sequence, being in any way smooth.

In the remainder of this chapter we consider four procedures:

1. fit a standard MLP with no weight decay penalty term – as the initial weights are selected close to zero, this may produce a set of weights that are serially correlated, much like the data;

2. smooth the weights after fitting the model;

[7]For a description of splines and spline models see Reinsch (1967); de Boor (1978) or Wahba (1990). For an overview see Smith (1979) or Wand (1999).

3. use a basis function expansion (Section 11.2.1 p. 185);

4. use the second difference penalty term (Section 11.2.2 p. 186).

We investigate these procedures in the following examples.

11.3   EXAMPLES

The examples considered in this section are:

- remotely sensed hyperspectral data from two instruments

  - the AVIRIS instrument (Section 11.3.2, p. 190); and

  - the HyMap instrument (Section 11.3.3, p. 193).

- data from a hand-held salinity scanner (Section 11.3.4) discussed by Kiiveri (1992);

11.3.1   Remotely sensed hyperspectral data

We consider the case of hyperspectral image data with radiance values for a large number of wavelengths. Being contiguous, these bands define a continuous spectrum of sufficient detail to differentiate the spectral signatures of many earth surface materials. This form of spectroscopy cannot be done with a small number of bands, as with Landsat TM's 7 bands.

The data sets considered are from two instruments: AVIRIS (Airborne Visible Infrared Imaging Spectrometer) and HyMap, both airborne[8]; we also consider spectral data from a ground-based, hand-held scanner. Figure 11.3 shows example spectra from each of the three instruments. It can be seen that the AVIRIS data has a number of noisy bands, due to atmospheric absorption. The HyMap instrument does not measure at these wavelengths and the hand held scanner appears less sensitive to this problem. However, in all cases it is necessary to drop a number of bands from the analysis.

Given the size of the images and the fact that the training sets should be of unambiguous class ascription, training sets are likely to contain fewer pixels than the number of spectral bands. In this case, the classification task presents certain difficulties, since:

- having more bands than observations gives rise to singular covariance matrices for canonical variate analysis and ML discrimination; and

- MLPs and other flexible discriminant techniques may achieve a spurious 100% correct classification on the training data, while having quite poor real error rates.

A common method of analyzing such images involves visual inspection and comparison with sample spectra from a laboratory-generated library, such as the U.S.

[8]The AVIRIS instrument is carried on a modified U2 spy plane at an altitude of approximately 20 km. The AVIRIS data are supplied by the Jet Propulsion Laboratory, Pasadena, California and are available for large areas of the United States, Canada and Europe. HyMap is carried on a conventional aircraft at much lower altitudes.

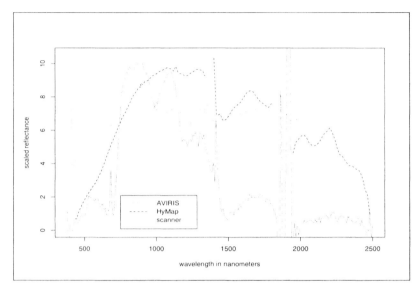

Figure 11.3 Three sample spectra from the three instruments considered. The AVIRIS spectrum is that of canola, the HyMap spectrum is a bare, heavily grazed, field and the hand-held scanner is a saline site. The HyMap data comes in three segments, avoiding the two regions where the Earth's atmosphere absorbs radiation. These spectra can be compared with the laboratory spectra in Figure 11.4.

Geological Survey digital spectral library. Figure 11.4 shows five spectra from this library: three minerals (calcite, carnallite and jarosite); and two samples of plant material (aspen and walnut leaf). The visible spectrum extends from 0.4 to 0.7 microns (approximately) and the bump visible at about 0.55 microns for the vegetation sample is the reflectance in the green band.

Inspecting Figure 11.4 it is apparent that there are large differences among the mineral spectra compared to the relative similarity of the two plant spectra. For mineral mapping, particular diagnostic troughs (absorption bands) in the spectra are isolated and compared.

The two plant spectra, however, show a similar shape that is characteristic of healthy green vegetation[9]. In addition, vegetation may have a variable spectral signature, depending, for example, on the season and weather conditions, and the spectral signature may follow a trajectory though spectral space (the "tessellated cap," see Richards, 1986) during the growing season. It is also possible that not all plants of a common species will be at the same point in the trajectory at identical times. For these reasons it appears unlikely that either measurements on a laboratory sample or a spectrum acquired in the field with a portable spectrometer will capture the full range of variation in a vegetation cover class.

So, on this evidence, we might expect that a different set of techniques may be necessary to distinguish ground-cover classes for vegetation than for minerals.

---

[9]Dry vegetation has a different spectrum.

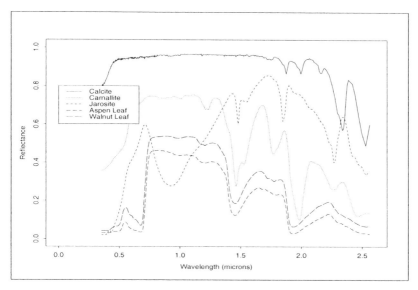

Figure 11.4   Five spectra from the USGS spectral library: three minerals (calcite, carnallite and jarosite) and two samples of plant material (aspen and walnut leaf). The spectra extend through the visible range and well out into the near and mid (thermal) infrared. The visible spectrum extends from 0.4 to 0.7 microns (approximately) and the bump visible at about 0.55 microns for the vegetation samples is the reflectance in the green band.

This brings us to a central point in the analysis that follows, and that is how the approach presented here differs from the standard analysis in the area of hyperspectral imagery. The standard approach is that of "unmixing" (Schowengerdt, 1997, §9). In this approach each pixel is considered to have a reflectance value that is determined by the fraction of several pure end-member materials contained in the pixel. The end-members are dependent on the image and are either determined from a library of laboratory-generated pure spectra or are determined from on-site spectral readings. It appears that the common practice is to use spectral libraries to get end-members for mineral mapping while vegetation mapping is done by acquiring spectra on site with a field spectrometer. We would argue that this shows that practitioners have found that spectral libraries do not capture the variation that is commonly seen in vegetation spectra[10].

If one adopts the unmixing approach, the gathering of ground data, if any, will consist of individual spectral readings of potential end-members. We have resorted to the preferred approach for MSS and TM data which has been to acquire ground truth covering a number of pixels. Often the ground truth is a map or image with regions of a common ground cover class drawn on the map by a farmer or

[10]Another unmixing approach (suggested in a personal communication to N. A. Campbell from R. O. Green (Jet Propulsion Laboratory, Pasadena, California) is to consider the chemistry and physics of the vegetation material. Pursuing this approach, laboratory spectra such as: chlorophyll a; chlorophyll b; beracarotene; cellulose; lignin; etc., are measured.

landholder. We can then select training areas from the known ground sites. For each ground cover class we thus have a representative range of spectra and an estimate of spectral variability within the ground cover class. The full question of the merits of each approach, while of great importance, is not pursued here[11].

11.3.2   AVIRIS data

AVIRIS delivers calibrated images of the spectral radiance from 0.4 to 2.5 microns (0.4 to 2.5 $\times 10^{-6}$ of a meter) in 224 contiguous spectral bands. The data can be visualized as an $n \times m \times 224$ block, commonly referred to as a "data cube."

The AVIRIS instrument has been used mainly for mineral and geological mapping, as opposed to ground cover classification, and consequently there does not appear to have been an organizational policy of actively acquiring ground truth. However, Roger Clark of the USGS was able to supply some information for a 512×614 image of an irrigated agricultural region of the San Luis Valley, Colorado. Unfortunately it transpired that what was initially taken to be a ground truth map was actually the result of a classification by a template fitting technique developed by Clark et al. (1991) with an unknown amount of ground verification. In response to questions about this the suppliers of the AVIRIS data recommended that we contact farmers directly to obtain ground truth; this has a large margin for error without a site visit and was not considered feasible.

Eight ground-cover classes were selected from the pseudo-ground-truth map[12]. The ground-cover classes, together with the number of sites and the total number of pixels, are given in Table 11.1.

Table 11.1   The eight training classes together with the number of sites and the total number of pixels per class.

class	sites	pixels
canola	6	897
potato	3	315
spinach	3	198
alfalfa 1	2	255
alfalfa 2	2	316
barley	7	977
chico	2	292
oat hay	2	272

---

[11]In some instances, unmixing methods may provide information that can not be provided by a classification based approach. In mineral exploration, having an estimate of the proportion of each end-member present in each pixel gives a view of the data that can not be produced with classification methods.

[12]The data and the ground truth have a number of imperfections:

- bands 1–13, 33, 97, 108–113, 154–176, 212–224 are corrupted and so the data was reduced to 167 spectral bands;

- the first training site (canola) was removed as it appeared quite atypical of the other canola sites;

- alfalfa was split into two ground cover classes, alfalfa 1 and alfalfa 2.

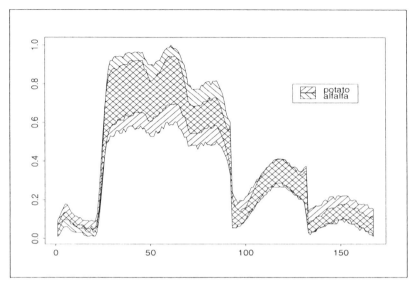

Figure 11.5 Normalized spectral response versus band number. The training area spectral envelopes for potato and alfalfa1 are overlayed. It is apparent that the potato and alfalfa spectra overlap throughout their range.

As discussed in Chapter 7 (p. 93) there is considerable correlation between the spectral values of neighboring pixels for TM data. This is likely to be the case for AVIRIS data as well. In addition, it may be the case that each training site is distinct from the other sites due to soil type, planting date, irrigation schedule etc. Because of this, the correct units of data to consider when doing cross–validation are the sites rather than the observations.

Visual inspection reveals that potato and alfalfa have a significant overlap throughout their spectral range (Figure 11.5) and so these two classes were chosen for a small-scale study. Each pair of sites (potato versus alfalfa 1) was used as training data in turn and tested on the remaining sites. The results, with a number of classifiers, are summarized in Table 11.2.

Linear discriminant analysis performs quite poorly and quadratic discriminant analysis fails as some of the training sites have singular covariance matrices. Nearest neighbor does somewhat better and the performance improves with the number of neighbors until $k = 55$.

The standard MLP gives an error of 13% and the use of a weight decay or weight elimination penalty causes an increase in the error rate. It is easy to find sets of weights that separate the training data; weight elimination (essentially a variable selection technique with only one hidden layer unit) appears to be using only two weights (see Figure 11.6). Due to the ill-posed nature of the problem, these sets of weights may not generalize well.

However, use of any of the suggested smoothing techniques causes a drop in error rates. The post-training smoothing was done with "super-smoother" (Friedman, 1984). The basis function MLP, which was fitted using a set of truncated Gaussians

Table 11.2   A comparison of several classification techniques on potato versus alfalfa classes. The test error is a cross-validated estimate across training sites.

classifier	test error rate
linear discriminant analysis	0.4504
penalized discriminant analysis ($\lambda = 0.2$)	0.3516
nearest neighbor (k=1)	0.2810
nearest neighbor (k=52)	0.2089
standard MLP	0.1274
standard MLP (weight elimination, ($\lambda = 0.00004$)	0.1309
standard MLP (weight decay, $\lambda = 1.2$)	0.1472
smoothed MLP	0.1182
second difference ($\lambda = 0.01$)	0.1121
basis MLP ($K = 25$)	0.0454

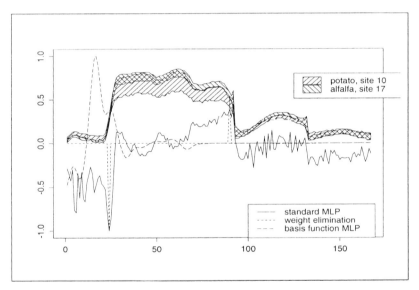

Figure 11.6   This plot combines an overlay of spectral site 10 (potato) and site 17 (alfalfa 1) with the weights from: a standard MLP; an MLP fitted with weight elimination ($\lambda = 0.00004$); and a basis function MLP (K=20). The weights are not shown at their true scale. As can be seen, weight elimination pushes all but two significant weights to 0.

with $\sigma = 2$ (ie two spectral bands) and 25 basis functions, gives an error rate of 4%.

The potato versus alfalfa1 discrimination task is one of the harder ones in this data set and the other separations are achieved readily with the same basis functions. Figure 11.7 shows the classified image achieved with the basis function MLP.

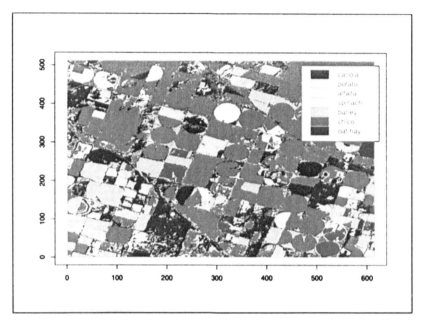

Figure 11.7   A section of the classified image for the San Luis Valley achieved with the basis function MLP.

11.3.3   HyMap data

The data consists of 105 bands (Figure 11.1, p. 181 shows a typical observation). This was the instrument's first field campaign (1996) and concerns have been raised about some aspects of the data[13]. Because of these concerns, these data are only used for illustrative purposes here.

A small test region (Figure 11.8) of size 125 by 125, was cut from the image. There are 14 training areas included in the the test region (described in Table 11.3) and an attempt was made to classify the region using each training area as a class. The classification was done using:

- a basis function MLP with a B-spline basis;

- flexible discriminant analysis (Hastie et al., 1995);

- smoothed discriminant analysis (Section 11.9, p. 183).

For this data set the gaps in the wavelengths are such that three separate sections must be fitted. For the basis function MLP with a B-spline basis, there were three knots at each of 1334.5 and 1790 nanometres (allowing the fitted function to have

---

[13]Professor Mervyn Lynch of Curtin University raised these concerns in a personal communication. Also I have raised the question of the informative nature of the noise in some bands. It appears that there is an "underflow" problem and that bands that should have a very small value, where the pixel contains water for example, end up with an essentially random value.

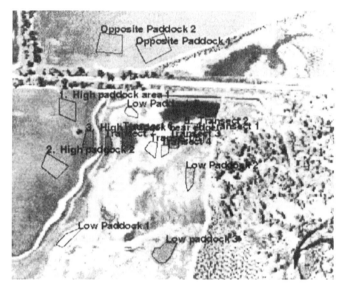

Figure 11.8   The training sites selected from the HyMap data for Toolibin for the first trial. The region shown is larger than the actual test region and serves to put it in context.

Table 11.3   The 14 training classes for the HyMap example and the number of pixels per class.

Low Paddock 5	55
Low paddock 3	173
Low Paddock 2	101
Low Paddock 1	96
Transect 7	60
Transect 6	40
Transect 5	45
Transect 4	22
Transect 3	31
Transect 2	60
Transect 1	48
High paddock near edge	88
High paddock 2	212
High paddock area 1	172

discontinuities there) and a total of 17 knots overall. Figure 11.9 shows the weights for the first hidden layer unit and Figure 11.10 shows the classified test region.

Similarly, the flexible and smoothed discriminant analyses required modifications to deal with the gaps in the wavelengths. For both of these methods a block diagonal matrix was used, with each block consisting of a D_2 matrix. The diagonal blocks

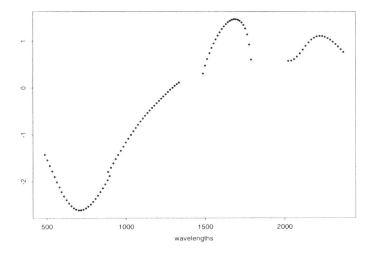

Figure 11.9 The weights fanning into the first hidden-layer unit from the basis-function MLP with a B-spline basis set.

were of sizes 59, 25 and 21 columns, respectively, so that

$$\underset{103\times105}{D_2} = \operatorname{diag}(D_2^{59}, D_2^{25}, D_2^{21})$$

and

$$\Omega = D_2^T D_2.$$

Figure 11.11 shows the image classified with a flexible discriminant analysis with $\lambda = 1$.

The smoothed LDA was fitted in the following manner. First we center the variables (columns of X). This makes calculating the covariance matrices easier but means that X is now not of full rank, hence the factor of $10^{-7}I$ below. Equation (11.9) (p. 183) is a generalized eigenvalue problem and can be written as

$$\Sigma_B \zeta = \theta \Sigma_W^\star \zeta \tag{11.12}$$

where θ is the eigenvalue and

$$\Sigma_W^\star = (1-\lambda)\Sigma_W + \lambda\Omega\frac{\operatorname{trace}(\Sigma_W)}{\operatorname{trace}(\Omega)} + 10^{-7}I$$

By a Cholesky (Strang, 1988) decomposition we find L such that $LL^T = \Sigma_W^\star$. The procedure is then:

- solve $LD = X^T$ for D;

- do a singular value decomposition to get $D = U\Lambda V^T$ and take the rotation matrix V;

- solve $L^T Z = V$ for Z. The canonical variates ζ are the columns of Z;

- normalize the Z so that $\zeta^T \Sigma_W \zeta = 1$.

Figure 11.12 shows the classified test region for $\lambda = 0.5$.

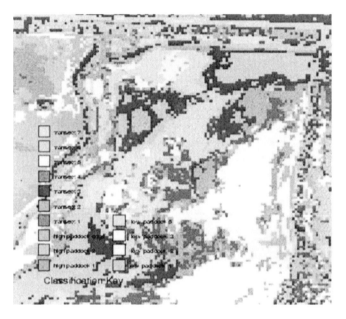

Figure 11.10   A classified image produced with a basis function MLP.

All three classification techniques have given plausible classified images. The speckled areas on the right-hand side of the images are due to the fact that there was no training class for bush.

As we noted before, this example is for illustration of the fitting methods only. It appears to be one of those cases where many methods can achieve very good accuracy on the training data and give quite different results on test data.

11.3.4   A hand-held salinity scanner

The scanning device measured 256 bands in the 0.4 to 2.5 micron range. The spacing between bands was 0.005 μm in the interval $(.402 - .684)$ μm, 0.008 in the interval $(.684 - 1.305)$ and 0.01 in the interval $(1.305 - 2.495)$ μm. The data consist of readings from 15 sites covering 4 different soil types.

Kiiveri (1992) provides further details and a canonical variate based approach to this classification problem. Figure 11.3 (p. 188) shows a typical spectrum from a saline site and shows a number of peaks that are in fact spurious features introduced

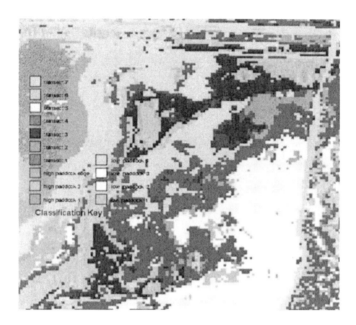

Figure 11.11   The image classified with a flexible discriminant analysis with $\lambda = 1$

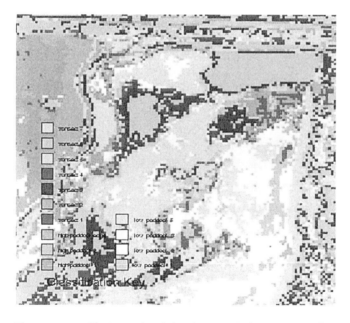

Figure 11.12   The image classified with a smooth discriminant analysis.

Table 11.4   Hand-held salinity scanner. "s" for salty, "n" for non-salty and "v" for variable.

site number	1	2	3	4	5	6	7	8	9	10	11	12	13	14	15
soil type	a	a	a	a	a	a	a	a	b	b	c	c	c	d	d
salinity status	n	n	s	s	n	s	v	v	n	n	v	n	s	s	n
sample size	9	10	10	11	10	10	10	10	15	22	15	11	13	14	17

by the scanner device. A consideration of atmospheric absorption regions and the scanner design led Kiiveri (1992) to limit the data to 175 bands, and we follow suit.

The test sites are described in Table 11.4. There are two trials reported here. For the first trial we only consider one of the soil types for which we have 80 pixels from a total of 8 sites. Of these sites, 3 have no salinity problems, 3 are saline, and 2 are classified as variable.

An attempt was made to analyze these data using a network with truncated Gaussian basis functions. A network of size 15.3.3 was reduced to 15.2.3 by task–based pruning (Chapter 6 p. 69). A clear picture emerges from the analysis, with saline and non–saline sites being well separated, and variable sites occupying an intermediate position (see Table 11.5). While a number of individual pixels (7) were misclassified, all sites were correctly classified (that is, correct on the majority of pixels). A plot of the weights for one of the two hidden-layer units is shown in Figure 11.13.

The second study omitted the variable class. Sites 4, 5, 10 and 13 were used for training data and sites 1, 2, 3, 6, 9, 12, 14, and 15 were used as test data. The study compared a number of classifiers with the results given in Table 11.6. The standard MLP of size 175.1.2 achieves the best performance on this problem.

Table 11.5   The confusion matrix for the three classes, non–saline soil, saline soil and soil of variable salinity. Note that the "variable" soil type overlaps with the other two but that there is no overlap between the non–saline and saline soils. This separation was achieved with an MLP using a truncated Gaussian set of basis functions of size 15.2.3.

soil type	non-saline	variable	saline
non-saline	28	1	
variable	1	14	5
saline			31

11.4   CONCLUSION

The AVIRIS example (see Figure 11.6) clearly shows that a variable selection technique may not be the method of choice for spectral classification. When the training data is limited relative to the number of bands, it may be the case that bands that will separate the training data will have little ability to separate independent test data. In this example, the smoothed MLP methods have performed very well.

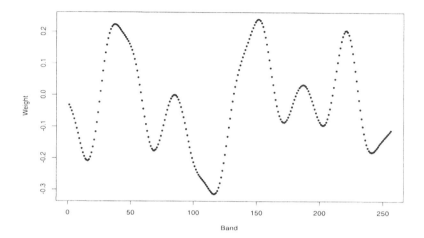

Figure 11.13   A plot of the weights for the hidden layer unit separating saline sites from sites of variable salinity.

Table 11.6   Comparison of several classification techniques on the scanner data. The error rates on the test and training data are given. Sites 4, 5, 10 and 13 were used for training data and sites 1, 2, 3, 6, 9, 12, 14, and 15 were used as test data.

classifier	training	test
Linear discriminant functions	0	0.2813
Nearest neighbor (k=1)	0	0.2187
standard MLP (175.1.2)	0	0.1459
standard MLP (weight elimination, $\lambda = 0.001$)	0.0577	0.1827
standard MLP (weight decay, $\lambda = 0.03$)	0	0.1771
smoothed MLP	0.0289	0.2115
second difference ($\lambda = 0.01$)	0.0865	0.1538
basis MLP ($K = 20$)	0.0673	0.2885
basis MLP ($K = 40$)	0.1731	0.2212
FDA ($\lambda = 0.0001$)	0.0673	0.2404

The HyMap example can not be assessed quantitatively due to the limited data. However, it appears that all three smoothed techniques are producing acceptable classifications in accordance with our knowledge of the Toolibin site.

Smoothing the weights after training gives an improvement in classification accuracy for most problems (but not the hand-held scanner). The second-difference penalty term and the basis function MLP both appear to give an improved performance over the standard MLP in most problems, however see Exercise 11.1 for a contrary case.

Overall, methods that model the data as sampled smoothed curves have a demonstrated utility in the classification of spectra. The presented examples demonstrate the utility of the two introduced approaches, alongside of other comparable methods such as penalized discriminant analysis (Hastie et al., 1995) and smoothed discriminant analysis (equation 11.9, p. 183).

11.5   COMPLEMENTS AND EXERCISES

11.1   This data set was used by Gorman and Sejnowski (1988) in their study of the classification of sonar signals using a neural network. The task is to train a network to discriminate between sonar signals bounced off a metal cylinder and those bounced off a roughly cylindrical rock.

The data set contains 111 patterns obtained by bouncing sonar signals off a metal cylinder at various angles and under various conditions and 97 patterns obtained from rocks under similar conditions. Each pattern consists of a vector of 60 readings produced by preprocessing the original sonar returns. See Gorman and Sejnowski (1988) for further details.

The experiment reported here is Gorman and Sejnowski's "aspect angle dependent experiment." The training and test sets are designed to ensure that both sets contain returns from each target aspect angle with representative frequency. Each set consists of 104 returns. Figure 11.14 shows a single example from each class, while Figure 11.15 shows the mean curves for the two classes. It appears that there is a significant degree of noise in the data.

The MLP models have very large numbers of parameters; for example a model with 12 hidden-layer units, as in the original study, has 758 parameters (including the bias terms) which exceeds the number of training examples[14]. In addition to these concerns about the number of parameters being fitted, there are good reasons for believing that in this case relatively few features of the data are actually required for a successful classification. Experienced human sonar operators report that they listen for three components in telling mines from rocks: the onset of the pulse (gradual or sudden); the strength and length of the central pulse; and the speed of the eventual decay of the signals. It seems implausible that human operators are using 60 features of the data to do the classification and yet their success rate varies from 88% to 97% (Gorman and Sejnowski, 1988)[15].

---

[14]However, as noted in Section 5.3.3 (p. 59), the effective degrees of freedom may be less than the number of parameters.

[15]The results with human operators were obtained in a different experiment and used the raw data. It may be the case that there is information in the raw data that is not preserved in the pre-processing.

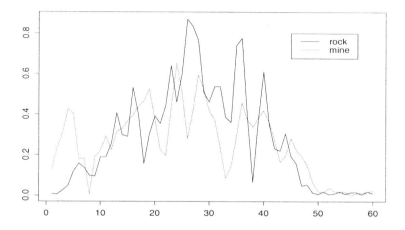

Figure 11.14   A single example of a mine and a rock sonar return. Variable value is plotted against variable number.

We were unable to replicate the results obtained by Gorman and Sejnowski and found that gradient descent minimization did not converge at all with the published step size and momentum term (Fahlman, 1990). In addition the nearest neighbor result on the test set published in Gorman and Sejnowski appears to be the 17.0% result for k=3 rather than the better 8% result for k=1.

We have used our own results (shown in Table 11.7) as a basis for comparison. The smoothed MLP methods have not performed well on the sonar-return data, failing considerably by comparison with the nearest neighbor results[16]. In addition the basis function MLP does not do better than the standard MLP on this data set.

11.2   Use the `rda` package to explore this data set.

11.3   The `smooth.lda` in the `mlp` library implements the penalized LDA procedure described in Section 11.1. Set up an Ω penalty matrix.

11.4   Generate a data set $X_i \sim N(\mu, \Sigma)$ there $\underset{P \times P}{\Sigma}$ shows correlations between components falling off as some function $f(|x_i - x_j|)$. Examine the eigenvectors of Σ.

---

[16] As in the STATLOG project (Mitchie et al., 1994), nearest neighbor methods have once again proved to be surprisingly good.

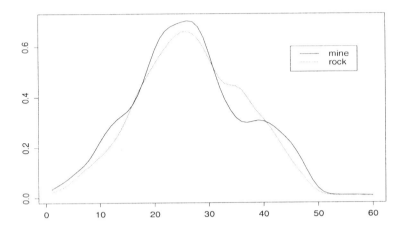

Figure 11.15   The mean curve for the mine and rock classes (found by smoothing splines). Variable value is plotted against variable number.

Table 11.7   A comparison of several classification techniques on the sonar returns data. The error rate is given for the training and test data sets.

classifier	training	test
Linear discriminant functions	0.0865	0.2404
Nearest neighbor (k=1)	0	0.0865
Nearest neighbor (k=2)	0	0.1635
Nearest neighbor (k=3)	0	0.1731
standard MLP (60.1.2)	0.0385	0.2019
standard MLP (weight decay, $\lambda = 0.19$)	0.1442	0.1538
standard MLP (weight elimination, $\lambda = 0.0$)	0.0385	0.2019
smoothed MLP	0.3077	0.1731
second difference ($\lambda = 0.01$)	0.0865	0.1538
basis MLP ($K = 20$)	0.0674	0.2885
basis MLP ($K = 40$)	0.1731	0.2212
FDA ($\lambda = 0.0001$)	0.0673	0.2404
FDA ($\lambda = 0.001$)	0.0962	0.2212
FDA ($\lambda = 0.01$)	0.1154	0.2115
FDA ($\lambda = 0.1$)	0.1538	0.2308
FDA ($\lambda = 1$)	0.1923	0.2404

CHAPTER 12

TRANSLATION INVARIANCE

12.1 INTRODUCTION

We consider the effects of certain linear pre-processing steps – centering and scaling – on the performance of an MLP model and uncover an intriguing bias.

Chapter 9 (p. 143) showed that the final solution of the MLP classifier is highly sensitive to the choice of the initial weights. Also it is clear from the experiments in Chapter 9 that the MLP classifier is not translation, rotation or scale invariant. In addition, we have shown that it is advantageous for the initial weights to be chosen with small magnitudes. However, there is no evidence in the literature that the choice of initial weights or data scaling introduces a consistent bias into the classifier.

The literature suggests that there is little compelling reason to rescale, standardize or normalize the data. Sarle (1996) suggests that there is no theoretical need to do so, as any rescaling of the input vector can be undone by a linear transformation of the corresponding weights. However, there may be practical reasons to do this, for example normalization may make training faster, as all variables will be on the same scale. In addition, it may reduce the chances of getting stuck in a local minimum.

Bishop (1995a, §9.2) discusses this question under the heading of "Consistency of weight decay". Bishop describes a consistency property of the MLP such that any linear transformation of the data will leave the network outputs unchanged if the inverse linear transformation is performed on the Ω weights. Thus the effect of

$$x_p \leftarrow ax_p + b \tag{12.1}$$

is undone by

$$
\begin{aligned}
\omega_{hp} &\leftarrow \frac{1}{a}\omega_{hp} \\
\omega_{h0} &\leftarrow \omega_{h0} - \frac{b}{a}\sum_p \omega_{hp}
\end{aligned}
\tag{12.2}
$$

(a similar property is discussed for transformations of the output variables). Bishop goes on to comment that

> If we train one network using the original data and one network using data for which the input and/or the target variables are transformed by one of the above linear transformations, then consistency requires that we should obtain equivalent networks which differ only by the linear transformation of the weights as given. Any regularizer should be consistent with this property, otherwise it arbitrarily favors one solution over another, equivalent one.

This then seems to summarize the prevailing state of knowledge about the question. There is no need to translate or scale the data, although such a transformation may be made for a variety of reasons including:

- reduced training time;

- avoiding undesirable local minima;

- if a penalty term is being used it is reasonable to scale all the variables to the same range. If the same penalty term is used on both the Υ and Ω weight matrices then the input variables should be scale to $[-1, 1]$, the same range as the outputs of the hidden layer.

However, even the standard, unregularized MLP is not consistent in Bishop's sense. To see why, we consider the penalty function[1] ρ and its derivative with respect to ω. ρ is defined on the joint feature and parameter spaces, and its derivative with respect to ω is also a function of both x and ω. Now, when we talk about the influence curve (i.e. the influence of an observation on the parameters(s) of interest) we are considering the ψ function (which has the same shape as the influence curve)

$$\psi(x, \omega) = \frac{\partial}{\partial \omega_{hp}}\rho(x, \omega)\bigg|_x$$

which is a function of x for fixed ω, so we are working in the feature space with the weights fixed. If, on the other hand, we are fitting the MLP model via some method that involves derivatives, then we are considering

$$s(x, \omega) = \frac{\partial}{\partial \omega_{hp}}\rho(x, \omega)\bigg|_\omega$$

---

[1] The arguments given here are not dependent on whether we use ρ_l (used in example three) or ρ_c (used in example two). The argument depends on the last two terms in the derivative of the ρ function, which are the same whether we use ρ_l or ρ_c.

EXAMPLE 1   **205**

so we are working in the parameter space with the data fixed[2]. Now in the case of minimizing the function ρ, when evaluating the gradient information at a particular point in weight space, both the weights and the data may be regarded as fixed (for the moment). Hence the evaluated derivative is simply the ψ function evaluated at each point in the training data and summed.

Consider the ψ function for ω_{hp} described in Section 8.5 (p. 128),

$$
\begin{aligned}
\psi_{\omega_{hp}} &= \left.\frac{\partial \rho}{\partial \omega_{hp}}\right|_{x} \\
&= \sum_{q=1}^{Q}\{t_q - \mathrm{mlp}_q(x)\}\frac{\partial f(z_q)}{\partial z_q}u_{q,h+1}\frac{\partial f(y_h)}{\partial y_h}x_p.
\end{aligned}
\tag{12.3}
$$

Say that we apply the linear transform (12.1) and the inverse transform (12.2). If we examine (12.3) we see that (12.1) and (12.2) will leave the term

$$
\frac{\partial f(y_h)}{\partial y_h}
$$

unchanged; however, the final x_i term will be affected by (12.1) but will be untouched by (12.2). This means that if (12.1) is applied to the weights of the MLP and (12.2) is applied to the training set, the output of the MLP will be unaffected, but the $\psi_{\omega_{hp}}$ functions will be different. As the $\psi_{\omega_{hp}}$ determine the behavior of a gradient–based minimization routine, after a minimization step the transformed and the untransformed MLP will have different weights.

Since the minimization step depends on the gradients, the next set of weights for the transformed and untransformed MLP will be different (and not recoverable one from the other by (12.2)). Practical experience (see below) indicates that the weights will continue to diverge.

Another way to view the problem is to consider where an observation has zero influence. Now for Fisher's Linear Discriminant Function (LDF), an additional point of class "+" has zero influence[3] if it is placed at the mean of class "+". For the MLP, we consider the Υ and the Ω weights separately. For the Υ weights, the IC is never zero for any finite value of x, although it may approach 0 asymptotically. For the Ω weights, $\psi_{\omega_{hp}}$ is 0 when x_p is 0, so that when the data are translated, the point of zero influence remains fixed at the point $x = 0$, not at the translated value. This introduces a consistent bias connected with linear transformations of the data.

12.2   EXAMPLE 1

To see the implications of this, we consider a simple problem with two classes and two features, x_1 and x_2. The classes are linearly separable and were chosen from uniform distributions on rectangular regions. This is the same data set as Experiment 2 from Section 9.3 (p. 145).

---

[2]If ρ is a log-likelihood then $s(x,\omega)$ is the Fisher score.
[3]This is true for the influence curve but only approximately true for the empirical influence curve. The additional point will change the sample size and thus the divisors of the covariance matrices and will also change the priors, if these are calculated from the number in each class.

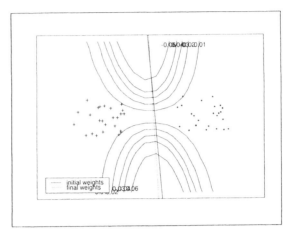

Figure 12.1   The initial decision boundary, chosen by the LDF, and the contours of the ψ function for the weight ω_2. Note the (approximate) symmetry of the contours about the x_2 axis; because of this a point above the data will have (approximately) an equal but opposite effect to a point of the same class symmetrically below the data. The data is approximately centered at $(0,0)$

We use the LDF to select the initial weights. A plot of the data set, with x_1 on the horizontal axis and x_2 on the vertical axis, is shown in Figure 12.1 along with the initial decision boundary, the ψ_{w_2} function at the initial weights as a contour plot, and the final decision boundary. We then shift the data by the addition of $(10, 10)$ to each data point, and we likewise shift the starting weights to

$$\omega_0 \leftarrow \omega_0 - 10\omega_1 - 10\omega_2,$$

so that the MLP is started from the corresponding position and has identical hidden-layer outputs and fitted values. The initial and final hyperplanes and the ψ function are illustrated in Figure 12.2 for a gradient based minimization. We can immediately see an interesting fact – shifting the data has caused the final decision boundary, using this minimization approach, to have a distinct slope.

As the data are drawn uniformly from two symmetric rectangular regions, they are roughly symmetric and hence we would expect some symmetry in $\psi(\omega_2)$. We see that the ψ function for the data centered at $(0,0)$ (Figure 12.1) has roughly symmetric contours with both positive (below the data) and negative (above the data) values. Hence expression (12.3) consists of a set of positive and negative numbers. If a point above the data moves the decision boundary in one direction, it would be expected that a point symmetrically below the data would have an equivalent but opposite effect.

However, when we shift the data set to $(10, 10)$ and shift the starting weights equivalently, the ψ values are now all negative and all exert an influence on ω_2 in the same direction as can be seen in Figure 12.2, which shows that the final fitted weights have a pronounced skew. Hence by shifting the data we have introduced a significant difference that manifests itself when we apply a minimization step in a gradient based minimization procedure.

EXAMPLE 1   207

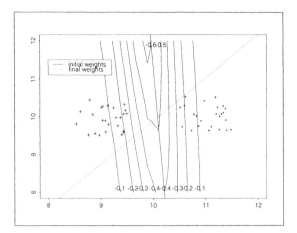

Figure 12.2   For the data set in Figure 12.1 shifted by $(10, 10)$, the plot shows the initial and final decision boundaries and the contours of the ψ function for the weight ω_2.

Where then would a data point have to be to exert an approximately equal influence in the opposite direction? It would have to be in the region of $(10, -10)$, as the $\psi(\omega_2)$ influence curve is not symmetric about the data, but about the x_1 axis.

This effect is not simply due to the fact that the two classes are well separated. In Figure 12.3 we have repeated the procedure for two overlapping classes, with similar results. Note that in this case the zero-centered data have an error rate of 0.21 and the data centered at (10,10) have an error rate of 0.24.

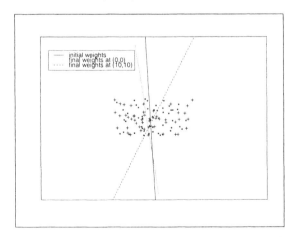

Figure 12.3   The initial and final decision boundaries for a non-linearly separable data set, first of all centered at the origin and then centered at $(10, 10)$. The translation of the data set has introduced a considerable bias into the fitting procedure.

12.3   EXAMPLE 2

A striking example of the bias imposed on the decision boundaries by linear trans-
lations of the data was afforded by an investigation of an approach for graphically
selecting initial decision boundaries. The method involves the following steps:

- the first two canonical variates (CVs) of the data are plotted - we will call
  this the first CV space;

- hyperplanes (lines) that separate groups of classes of interest are selected
  visually. (This involves clicking a pointer on the CV plot in *S–PLUS* .) The
  selected lines are returned in a matrix at the end of the graphical selection
  process;

- the separated classes are removed from the data and the remaining data are
  used to calculate a new CV space. This new CV space is a transformation
  of the first two-dimensional CV space and not of the original data. The
  remaining data are plotted in the new CV space and the process is repeated;

- all the selected hyperplanes are transformed to the first CV space and are
  used as starting values for an MLP model.

The method was applied to data derived from 6-band Landsat Thematic Mapper
images of a region in the wheatbelt of Western Australia. Ten classes were present
in the data: water; swamp; bush; bare; salt 1; salt 2; poor 1; poor 2; poor 3; and
average. The aim here was to identify ground that had been adversely affected by
salinity.

This example has some interest in its own right. In the space of the 6 variables,
the classes of interest (water, swap etc are not really of interest in this study) do
not form separate clumps. As can be seen from Figure 12.4, "salinity" appears to
be part of a central mass of pixels that includes the "poor" and "bush" classes.

Using LDA to classify data of this sort is a generally an interactive and iterative
process particularly if there is no ground truth available. It involves selecting
regions that are unambiguously of a class then selecting other regions, seeing where
they lie in canonical variate space and perhaps adding them to the training data.
Automated tools to assist this labor intensive process are quite desirable.

However, our use of the example occurs after the initial weights have been selected. Four trials are reported:

- one with the raw data, shown in Figures 12.4 12.5;

- one with the data mean-centered, shown in Figure 12.6;

- one with the data scaled to the interval $[-1, 1]$, shown in Figure 12.7; and

- one with data centered and scaled to the interval $[-1, 1]$, shown in Figure
  12.8.

The initial hyperplanes and decision boundaries, as shown in Figure 12.4, are the
same, relative to the data, in each case. The only thing that differs in these three
trials is the scaling and centering of the data. Clearly the linear transformation of
the data has had a major affect on the final decision regions.

EXAMPLE 3   209

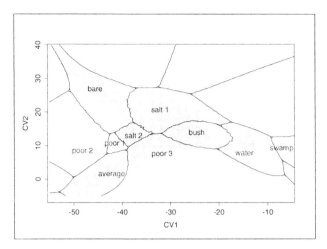

Figure 12.4   The initial decision regions using the raw (unscaled and uncentered) data.

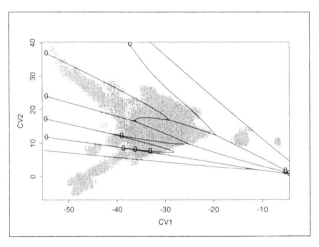

Figure 12.5   The final decision regions using the raw data. From the shape of the decision regions it can be seen that all of the hyperplanes are roughly pointed toward the origin.

12.4   EXAMPLE 3

Initial comments[4] on Dunne (2000) included the following:

- when the classes are separable, it is possible to achieve a zero error if the output of the MLP is a Heaviside step function, and in this case an infinite number of solutions are possible, all giving zero error. When one of these

[4]The comments were by members of the Remote Sensing group, CSIRO

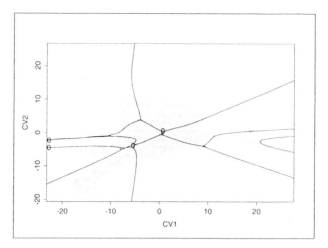

Figure 12.6   The final decision regions using the centered data. Note that the decision regions on the right of the image still appear to be formed from hyperplanes which are roughly pointed toward the origin.

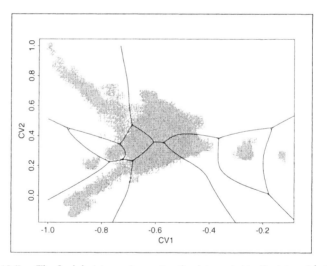

Figure 12.7   The final decision regions using the data scaled to the interval $[-1, 1]$,

solutions is attained, there should then be no biasing of the MLP toward solutions that pass through the origin;

- perhaps the bias is an artifact of the minimization routine and a non-derivative minimization routine such as the simplex algorithm might not be subject to the same bias.

EXAMPLE 3   **211**

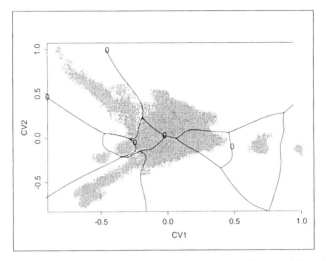

Figure 12.8   The final decision regions using the data centered and scaled to the interval $[-1, 1]$.

The first point is correct, if the MLP achieves a zero error and a Heaviside step output it will not move to a biased solution. However, the Heaviside step function is approached asymptotically by the MLP and for any finite values of the weights, the MLP will have a continuous and differentiable output and will behave as outlined above. As argued in Chapter 9 (p. 143), starting with weights of small magnitude is essential to finding an acceptable local minimum. It is only in a trivial case like Example 1, where we know the solutions before we train the MLP, that we could start with weights of large magnitude and annul the effects of the bias. In this case our final decision boundary will be very similar to our initial decision boundary. So it is only in the case where we know the correct answer and can thus afford to start the MLP with initial weights of extremely large magnitude (giving it a Heaviside step function output) that we can avoid the consistent bias described here.

The second point is false. The following figures show a data cloud situated at the points $(0,0)$, $(10,10)$, $(10,-10)$, $(-10,10)$, and $(-10,-10)$. At each location, the initial and final decision boundaries are shown for an MLP model. As can be seen for quasi-Newton (Figure 12.9), simplex (Figure 12.10) and conjugate gradients (Figure 12.11) minimization routines, the MLP models have a high bias for all examples except those situated at $(0,0)$. The linear transformations have a genuine effect on the shape of the error surface and any minimization algorithm that moves in a locally downhill direction from the starting values will be affected.

We have shown that the derivatives ensure that any gradient-based algorithm will tend towards the oblique solution. However, it might reasonably be argued that for a given slope of the activation function, a vertical orientation must give a lower ρ value than an oblique orientation, as illustrated in Figure 12.12. That is, the global minimum is still given by the same decision boundary with a vertical

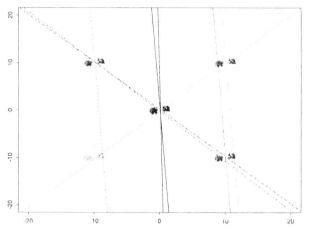

Figure 12.9   The figure shows data clouds situated at the points $(0,0)$, $(10,10)$, $(10,-10)$, $(-10,10)$, and $(-10,-10)$. At each location the initial and final decision boundaries are shown for an MLP model. The lines intersecting between the two classes are always the initial and final decision boundaries for those classes. In addition, the initial and final lines for each data set are in the same continuous/dashed line style. For a quasi-Newton minimization procedure, the MLP models have a high bias for all examples except those situated at $(0,0)$.

orientation, it is only the local minima that are subject to the bias. However, the global minimum is now effectively denied to a descent algorithm.

However, recall that the simplex algorithm (see Appendix A.10, p. 254) has the capacity to leave the region of attraction of a local minimum and search more widely in the feature space. In order to test if the simplex algorithm can jump to a good, but not locally downhill, solution in this situation, a series of further trials were undertaken.

Figure 12.13 shows the iterations of the simplex algorithm as it moves to its final position. When these lines are plotted as the algorithm is running, it can be seen that the simplex algorithm, unlike the other algorithms, does not move to its final position smoothly but instead jumps back and forth. Figures 12.14 and 12.15 show the iterations from starting values that are respectively: the original starting values multiplied by 10; and the original starting values divided by 10. Clearly increasing the weights has reduced the tendency of the MLP to move to the oblique position.

The finite precision of computer arithmetic means that effective infinity is reached when the difference between the sigmoid activation function and the Heaviside step function is less than the machine precision. When this region is reached the output of the MLP is effectively a step function and any of an infinite number of solutions will give a zero error. From a given starting position in weight space both:

- an increase in the magnitude of the weights; and

EXAMPLE 3   **213**

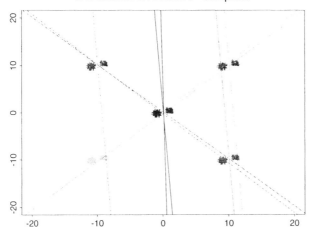

Figure 12.10   The figure shows data clouds situated at the points $(0,0)$, $(10,10)$, $(10,-10)$, $(-10,10)$, and $(-10,-10)$. At each location the initial and final decision boundaries are shown for an MLP model. The lines intersecting between the two classes are always the initial and final decision boundaries for those classes. In addition, the initial and final lines for each data set are in the same continuous/dashed line style. For a simplex minimization procedure, the MLP models have a high bias for all examples except those situated at $(0,0)$.

- a movement of the weights to put the decision boundary through the origin,

both cause the ρ value to drop. However, for a gradient-descent algorithm, the move towards zero will predominate as this is the direction of steepest descent. Simplex, on the other hand, may jump directly towards the region at infinity. Now as the magnitude of the weights increases ρ becomes "stuck" as the derivative is close to 0 (bringing gradient based minimization schemes to a halt) and ρ changes little per unit change in the weights. Figure 12.16 shows this diagrammatically.

An interesting question is whether the simplex algorithm can jump to the point at effective infinity from the given starting positions. The answer is that it can if the initial step size is set large enough[5]. The algorithm moves from initial weights of

$$\Upsilon = \begin{pmatrix} -1.098 & 2.208 \\ 1.098 & -2.208 \end{pmatrix}$$

to

$$\Upsilon = \begin{pmatrix} -128.255 & 198.221 \\ 77.224 & -228.463 \end{pmatrix}$$

at which point the outputs of the MLP are 0 and 1 to machine precision. Remarkably the Ω weights were not changed at all, remaining at

$$\Omega = \begin{pmatrix} -37.237 & 3.477 & 0.217 \end{pmatrix}.$$

[5] In this example it was set to 100 times its default value of 1.

minimization routine -- conjugate gradients

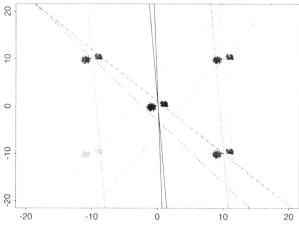

Figure 12.11   The figure shows data clouds situated at the points $(0,0)$, $(10,10)$, $(10,-10)$, $(-10,10)$, and $(-10,-10)$. At each location the initial and final decision boundaries are shown for an MLP model. The lines intersecting between the two classes are always the initial and final decision boundaries for those classes. In addition, the initial and final lines for each data set are in the same continuous/dashed line style. For a conjugate gradients minimization procedure, the MLP models have a high bias for all examples except those situated at $(0,0)$.

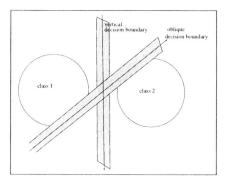

Figure 12.12   For a given slope of the activation function, the vertical orientation must give a lower ρ value than the oblique orientation.

12.5   CONCLUSION

It is clear from an examination of the shape of the ψ function, and from these examples, that if the data are not translated to be "near" the origin then a significant bias may be introduced into the MLP classifier. It is clear that both centering the

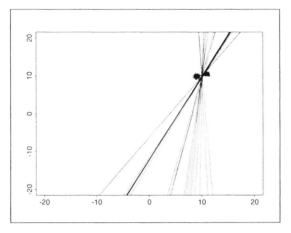

Figure 12.13   The figure shows the iterations of the simplex algorithm as it moves to its final position as shown in Figure 12.10.

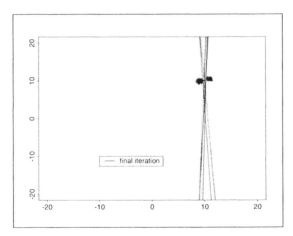

Figure 12.14   The figure shows the iterations of the simplex algorithm as it moves to its final position. The starting values were the starting values for Figure 12.13 multiplied by 10.

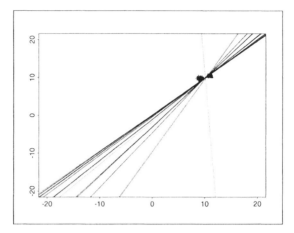

Figure 12.15   The figure shows the iterations of the simplex algorithm as it moves to its final position. The starting values were the starting values for Figure 12.13 divided by 10.

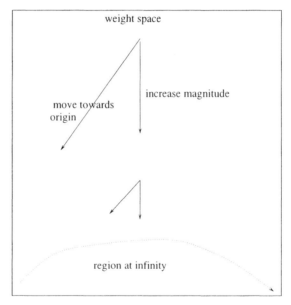

Figure 12.16   As the magnitude of the weights increases the difference in ρ for the two strategies: increasing the magnitude of the weights; and moving the decision boundary close to the origin, becomes less pronounced. When the weights are effectively infinite there is no difference between these strategies.

data and scaling to moderate maximum values both have a beneficial effect on the final decision boundaries.

The strategies arising from this investigation are obvious. What is more surprising is how this has escaped notice in the literature. We offer two possible reasons for this. One is than MLPs have very frequently been tested on problems much harder than the test example presented here. For hard high dimensional problems the resulting bias may be hard to see. In addition the flexible nature of the MLP classifier may mean that for some problems the results achieved with an MLP classifier are usable despite the bias.

The other is that the bias can be avoided by practice of normalizing the columns of X so that each variable is in the range $[-1, 1]$. This is advisable if one is using a single parameter λ with, say, a weight decay penalty term, and ensures that all weights in the MLP (on all layers) have a multiplicand in the same range. The use of penalty terms and this scaling has undoubtedly insulated some fitted MLP models from the bias described here.

CHAPTER 13

FIXED-SLOPE TRAINING

13.1 INTRODUCTION

We have seen various strategies for penalized fitting in MLP models summarized in Section 5.4 (p. 62) and examined some strategies for making MLP models more robust. We now take up the work of Bartlett (1998) who suggests that the magnitude of the weights is the most important consideration in avoiding overfitting, more important than the size of the network. To address this directly we consider a strategy for fixing the magnitude of the L^2 norm of the weight vectors during the fitting process[1].

As shown in Chapter 9 (p. 143) during the fitting process of the MLP, we go from the initial weights which are selected from a zero-centered region of small magnitude to larger weights. This causes two things to happen:

1. individual points close to the decision boundary have an increasing "gross error sensitivity," (Section 8.4.3, p. 125). This may lead to single points moving the decision boundary in undesirable ways; and

2. the decision boundary between classes may become increasingly ragged (depending on the number of hidden-layer units).

---

[1] Bartlett (1998) actually considers bounds on the L^1 norm of the weights. However it is much easier to bound the L^2 norm.

Both are aspects of the same problem: we are using a very flexible method that can model noise as well as structure in the data. In order to overcome this we introduce a form of penalized training called "fixed-slope training" in which the length of the weight vector fanning into a unit is fixed or constrained.

This differs from the weight decay (ridge regression, p. 64) penalized term, $\lambda \sum_r \omega_r^2$, in that it is implemented so as to separate the magnitude of the weights (which determines the slope of the logistic function) from the position of the decision boundary. As noted in Tibshirani (1996) ridge regression has a tendency to make the parameters equal to minimize the squared norm and so applying the penalty $\lambda \sum_r \omega_r^2$ may cause some parameters to increase. Tibshirani (1996) gives an example of this.

"Fixed-slope training" also differs from the "least absolute shrinkage and selection operator" (LASSO) discussed in Tibshirani (1996) and Osborne et al. (1998). This puts a penalty on the sum of the absolute values of the weights, namely, $\lambda \sum_r |\omega_r|$. See also the discussion in Section 5.4, p. 62.

The proposed procedure fixes the slope of the activation function, prevents large weights, limits the relative influence that individual points have on the final weight values, and thus smoothes the decision boundaries[2]. To allow fixed slope training to be implemented we transform the weights to polar coordinates. We take the weights $\{w_p\}_{p=1}^{P+1}$ fanning into a unit to be the Cartesian coordinates of a point in weight space and re-parameterize them in radial coordinates. If

$$\Theta = (\gamma, \theta_1, \ldots, \theta_P)$$

is a point in radial coordinates, then the transformation is defined by the $P + 1$ mappings

$$w_{P+1} = \gamma \sin(\theta_1)$$
$$w_P = \gamma \cos(\theta_1) \sin(\theta_2)$$

$$\vdots$$

$$w_2 = \gamma \cos(\theta_1) \cos(\theta_2) \ldots \cos(\theta_{P-1}) \sin(\theta_P)$$
$$w_1 = \gamma \cos(\theta_1) \cos(\theta_2) \ldots \cos(\theta_{P-1}) \cos(\theta_P),$$

subject to the constraints that

$$\gamma \geq 0$$
$$\theta_1 \in \left[-\frac{\pi}{2}, \frac{\pi}{2} \right]$$

and

$$\theta_p \in [0, 2\pi] \text{ for } p = 2, \ldots, P.$$

We can then make Θ the adjustable parameters of the unit. To fit the model we need to calculate

$$\frac{\partial \rho}{\partial \gamma} \text{ and } \frac{\partial \rho}{\partial \theta_p} \text{ for } p = 1, \ldots, P.$$

---

[2] If we consider the weights associated with a unit (rather than the whole weight vector) fixed-slope training has a rather weak Bayesian interpretation as a prior of uniform distribution on a sphere of fixed radius in weight space.

Consider ω_h, the weights fanning into hidden–layer unit h, and Θ_h, the same weights in polar coordinates. Then

$$
\left[\frac{\partial \omega_h}{\partial \Theta_h} \right] =
\begin{bmatrix}
\dfrac{\partial \omega_1}{\partial \gamma} & \dfrac{\partial \omega_2}{\partial \gamma} & \cdots & \dfrac{\partial \omega_{P+1}}{\partial \gamma} \\
\vdots & \vdots & \vdots & \\
\dfrac{\partial \omega_1}{\partial \theta_P} & & \cdots\cdots\cdots & \dfrac{\partial \omega_{P+1}}{\partial \theta_P}
\end{bmatrix}
$$

a $P + 1 \times P + 1$ matrix (suppressing the index h),

$$
=
\begin{bmatrix}
\dfrac{\omega_1}{\gamma} & \dfrac{\omega_2}{\gamma} & \cdots\cdots\cdots & \cdots\cdots\cdots & \dfrac{\omega_{P+1}}{\gamma} \\
\dfrac{-\omega_1 \sin(\theta_1)}{\cos(\theta_1)} & \cdots\cdots\cdots & & \dfrac{-\omega_P \sin(\theta_1)}{\cos(\theta_1)} & \dfrac{\omega_{P+1}\cos(\theta_1)}{\sin(\theta_1)} \\
\dfrac{-\omega_1 \sin(\theta_2)}{\cos(\theta_2)} & \cdots & & \dfrac{-\omega_{P-1}\sin(\theta_2)}{\cos(\theta_2)} \quad \dfrac{\omega_P \cos(\theta_2)}{\sin(\theta_2)} & 0 \\
\vdots & & & \vdots & \vdots \\
\dfrac{-\omega_1 \sin(\theta_P)}{\cos(\theta_P)} & \dfrac{\omega_2 \cos(\theta_P)}{\sin(\theta_P)} & \cdots & 0 & 0
\end{bmatrix}.
$$

Then

$$
\frac{\partial \rho}{\partial \Theta_{qh}} = \frac{\partial \rho}{\partial z^*}\frac{\partial z^*}{\partial z_q}\left(\left[\frac{\partial v_q}{\partial \Theta_q} \right] y^* \right)[h],
$$

where

$$
\left[\frac{\partial v_q}{\partial \Theta_q} \right] y^*
$$

is a vector of length $H + 1$ and we take the hth element of this vector. For $h = 1, \ldots, H$ and $p = 1, \ldots, P + 1$ we have

$$
\frac{\partial \rho}{\partial \Theta_{hp}} = \sum_{q=1}^{Q} \frac{\partial \rho}{\partial z^*}\frac{\partial z^*}{\partial z_q} v_{q,h+1}\frac{\partial f(y_h)}{\partial y_h}\left(\left[\frac{\partial \omega_h}{\partial \Theta_h} \right] x \right)[p].
$$

So by comparison with the derivatives for the standard MLP, the change to polar coordinates will require an extra inner-product in each iteration.

13.2 STRATEGIES

Now that we have the mechanism to fit an MLP with radial weights, there are a number of strategies that we can follow including:

- let Θ vary without restraint – the MLP then gives essentially the same solution as in the standard case;

- add a penalty term of the form

$$
\rho_p = \rho + \lambda \sum_{\text{all } \gamma \text{ terms}} \gamma;
$$

- constrain γ such that $\gamma \leq \lambda$ for some λ; this is done by setting $\dfrac{\partial \rho}{\partial \gamma} = 0$ and $\gamma = \lambda$ when γ is greater than λ;

- fix γ and let the θ vary;

- fix γ and let the θ vary and then fix θ and let the γ vary. This gives the same separating boundary as the previous approach, but the probability estimates will differ;

- apply the fixed-slope constraint to the Υ weights, the Ω weights or to both sets of weights.

The standard approach that we have investigated is to fix γ to the same value for both sets of weights.

13.3   FIXING Υ OR Ω

While it is possible to apply the fixed-slope constraint to the Υ weights, the Ω weights or to both sets of weights, somewhat surprisingly it is in applying the constraint to the Υ weights that the MLP model becomes more resistant. This can be seen by considering the geometry of the problem. The Ω weights are needed to determine the position of a decision hyperplane; the actual numerical values of the outputs of the hidden layer are not relevant (provided one super-class gives values below α and the other super-class above α, the actual value of α is not relevant). The Υ weights determine the combination of the hyperplanes for decision regions and also attempt to minimize the penalty function. In order to do this the outputs of the final layer must approach as close to 0 and 1 as possible, and hence fixing the slope constrains the Υ weights and provides resistance to aberrant data points.

13.4   EXAMPLE 1

We consider the data set from Experiment 2 in Section 9.3.2 (p. 146). Here, as in Chapter 9, a single point has been moved and the resulting effects noted on the decision boundaries, which are calculated for the linear discriminant function (LDF), a standard MLP model and a fixed-slope MLP.

Figure 13.1 shows the effects of fixing either the Ω or the Υ weights alone.

An inspection of Figure 13.2 shows that neither the LDF nor the standard MLP are resistant to an aberrant data point, although their non-resistant behavior is quite different. The LDF is influenced by a point in class "+" by virtue of how far that point is (in Mahalanobis distance) from the mean of class "+," whereas the standard MLP has a non-linear influence function that has a pronounced ridge along the lines of linear separability. This difference is evident in plots 4 and 5, where a small movement of the point (across the bounds of linear separability) makes a considerable change in the behavior of the standard MLP, while leaving the LDF virtually unchanged.

In all of the plots it is clear that the fixed-slope MLP is resistant to the aberrant point. In addition, by comparison with the sensitivity plots for the standard MLP on the data set the sensitivity curve for the fixed-slope MLP is very flat.

EXAMPLE 2   **223**

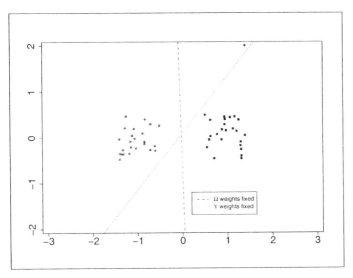

Figure 13.1   The effects of fixing the Ω weights alone versus the effects of fixing the Υ weights alone. It is fixing the Υ weights that confers a robustness property on the MLP.

13.5 EXAMPLE 2

Example 2 is the simulated spatially correlated data set considered in Section 7.1 (p. 114). A simulated scene was constructed containing three classes (bush, wheat and pasture) and four spectral bands for each pixel (see Figure 7.9, p. 114). This constituted the test data. We generated three 15×15 training areas (bush, wheat and pasture) using the same process.

We consider the problem without using neighbor information. Table 13.1 gives the error rates for several classifiers.

Table 13.1 replicates values given in Tables 7.2 7.3, and 7.4. These were not the best results achieved on this data set (see Section 7.1, p. 114). Clearly the penalized solutions, weight decay and fixed slope, are better than the non-penalized solutions. The differences between the non-penalized and the penalized solutions are significant, while the differences between the penalized solutions are barely noticeable in comparison.

13.6 DISCUSSION

It is clear that in ill-posed problems, such as the given example, the need for a penalization term is paramount.

Fixed-slope training appears to work well as a penalization method. While its implementation is more complex than that of weight decay, for example, it has an intuitive geometric interpretation in terms of the slopes of the activation functions, and of the fitted surface. Fixing the slope of the activation function makes the MLP model more resistant to the influence of isolated points.

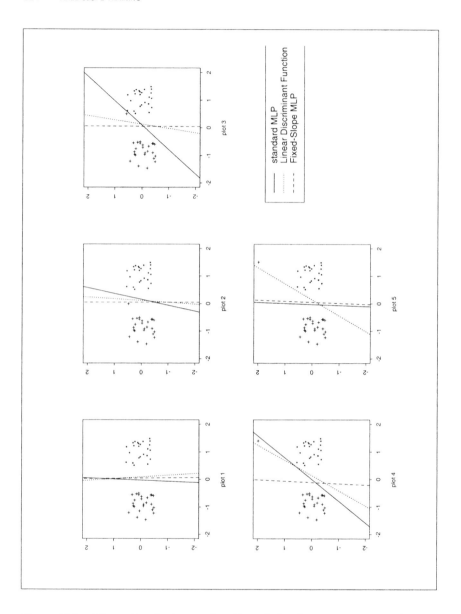

Figure 13.2 We consider the data set from Example 2 in Section 9.3.2, p. 146. As in Chapter 9, a single point has been moved and the resulting effects noted on the decision boundaries, which are calculated for the linear discriminant function (LDF), a standard MLP model and a fixed-slope MLP. An inspection of the figure shows that neither the LDF nor the standard MLP are resistant to an aberrant data point, although their non-resistant behavior is quite different. The LDF is influenced by a point in class "+" by virtue of how far that point is (in Mahalanobis distance) from the mean of class "+," whereas the standard MLP has a non linear influence function that has a pronounced ridge along the lines of linear separability. See particularly the difference between plots 4 and 5. The fixed slope MLP is resistant to the moving point.

Table 13.1   A comparison of several classification techniques on the simulated scene. For each classifier the error rate on the image is given.

classifier	error rate
Linear discriminant functions	0.2536
Quadratic discriminant functions	0.2552
Nearest neighbor	0.2320
MLP (of size 4.2.3), a cross–entropy penalty function, softmax outputs	0.2600
MLP, as above, with weight decay $\lambda = 0.1$	0.2216
MLP, as above, with fixed slope, unconstrained	0.2776
MLP, as above, with fixed slope $\gamma = 6$	0.2208
MLP, as above, with fixed slope $\gamma = 5$	0.2192
MLP, as above, with fixed slope $\gamma = 4$	0.2224
MLP, as above, with fixed slope $\gamma = 3$	0.2312

Bibliography

Adby, P. R. and Dempster, M. A. H. (1974). *Introduction to Optimization Methods.* Chapman and Hall, London.

Ahmed, S. and Tesauro, G. (1988). Scaling and generalization in neural networks: a case study. In Touretzky et al. (1988), pp. 3–9.

Aires, F., Prigent, C., and Rossow, W. B. (2004). Neural network uncertainty assessment using bayesian statistics: A remote sensing application. *Neural Computation*, 16(11):2415–2458.

Aitchison, J. and Dunsmore, I. R. (1975). *Statistical Prediction Analysis.* Cambridge University Press, Cambridge.

Almeida, L. and Wellekens, C., editors (1990). *Lecture Notes in Computer Science 412, Neural Networks EURASIP Workshop Sesimbra, Portugal.* Springer–Verlag, Berlin.

Alpaydin, E. (1991). Gal: Networks that grow when they learn and shrink when they forget. Preprint, International Computer Science Institute, Berkeley, CA.

Amari, S., Murata, N., Müller, K.-R., Finke, M., and Yang, H. (1997). Asymptotic statistical theory of overtraining and cross-validation. *IEEE Transactions on Neural Networks*, 8(5). Previously, University of Tokyo Techreport: METR 95-06, 1995.

Anderson, J. A. and Rosenfeld, E., editors (1988). *Neurocomputing: Foundations of Research.* MIT Press, Cambridge, MA.

Anderson, T. W. (1984). *An Introduction to Multivariate Statistical Analysis.* Wiley, New York, second edition.

Anderson, T. W. and Bahadur, R. R. (1962). Classification into two multivariate normal distributions with different covariance matrices. *Annals of Mathematical Statistics*, 33:420–431.

Apostol, T. (1967). *Calculus, Vol. I & Vol II.* Wiley, New York.

Baldi, P. (1991). Computing with arrays of bell-shaped and sigmoid functions. In Lippmann et al. (1991), pp. 735–742.

Baldi, P. and Hornik, K. (1995). Learning in linear neural networks: A survey. *IEEE Transactions on Neural Networks*, 6(4):837–858.

Barnard, E. and Cole, R. A. (1989). A neural–net training program based on conjugate–gradient optimization. Technical Report CSE 89–014, Oregon Graduate Institute of Science and Technology.

Barron, A. R. (1984). Predicted squared error: A criterion for automatic model selection. In Farlow, S. J., editor, *Self-organizing Methods in Modeling.* Marcel Decker, New York.

Bartlett, P. L. (1998). The sample complexity of pattern classification with neural networks: The size of the weights is more important than the size of the network. *IEEE Transactions on Information Theory*, 44(2):525–536.

Barton, S. A. (1991). A matrix method for optimizing a neural network. *Neural Computation*, 3:450–459.

Battiti, R. (1992). First– and second–order methods for learning: between steepest descent and Newton's method. *Neural Computation*, 4:141–166.

Baum, E. and Haussler, D. (1989). What size net gives valid generalization? *Neural Computation*, 1:151–160.

Baum, E. B. (1990). When are k–nearest neighbor and back propagation accurate for feasible sized sets of examples? In Almeida and Wellekens (1990), pp. 2–25.

Becker, R. A., Chambers, J. M., and Wilks, A. R. (1988). *The new S Language.* Wadsworth and Brooks, Monteray, CA.

Becker, S. and le Cun, Y. (1989). Improving the convergence of back-propagation learning with second order methods. In Touretzky et al. (1988), pp. 29–37.

Benediktsson, J. A., Sveinsson, J. R., and Arnason, K. (1995). Classification and feature extraction of AVIRIS data. *IEEE Transactions on Geoscience and Remote Sensing*, 33(5):1194–1205.

Besag, J. (1974). Spatial interaction and the statistical analysis of lattice systems (with discussion). *Journal of the Royal Statistical Society B*, 36(2):192–236.

Besag, J. (1986). On the statistical analysis of dirty pictures (with discussion). *Journal of the Royal Statistical Society B*, 48(3):259–302.

Billingsley, P. (1995). *Probability and Measure*. Wiley, New York, 3rd edition.

Birch, M. W. (1963). Maximum likelihood in three-way contingency tables. *Journal of the Royal Statistical Society B*, 25:220–233.

Bischof, H., Schneider, W., and Pinz, A. (1992). Multispectral classification of Landsat images using neural networks. *IEEE Transactions on Geoscience and Remote Sensing*, 30(3):482–490.

Bishop, C. M. (1993). Curvature-driven smoothing: a learning algorithm for feed-forward networks. *IEEE Transactions on Neural Networks*, 4:882–884.

Bishop, C. M. (1995a). *Neural Networks for Pattern Recognition*. Oxford University Press, Oxford.

Bishop, C. M. (1995b). Training with noise is equivalent to Tikohonov regularization. *Neural Computation*, 7:108–116.

Breiman, L. (2001). Statistical modelling: The two cultures (with discussion). *Statistical Science*, 16(3):199–231.

Breiman, L., Friedman, J. H., Olshen, R. A., and Stone, C. J. (1984). *Classification and Regression Trees*. Wadsworth and Brooks/Cole, Monterey, CA.

Breiman, L. and Ihaka, R. (1984). Nonlinear discriminant analysis via ACE and scaling. Technical Report 40, Dept. of Statistics, University of California, Berkeley.

Bridle, J. S. (1990). Training stochastic model recognition algorithms as networks can lead to maximum mutual information estimation of parameters. In Touretzky (1990), pp. 211–217.

Brown, M. P. S., Grundy, W. N., Lin, D., Cristianini, N., Sugnet, C., Furey, T. S., Manuel Ares, J., and Haussler, D. (2000). Knowledge-based analysis of microarray gene expression data using support vector machines. *Proceedings of the National Academy of Sciences*, 97(1):62–267.

Buja, A., Hastie, T., and Tibshirani, R. (1987). Linear smoothers and additive models. Statistical Research Reports 50, AT&T Bell Laboratories.

Burges, C. J. C. (1998). A tutorial on support vector machines for pattern recognition. *Data Mining and Knowledge Discovery*, 2(1):121–167.

Campbell, N. A. (1978). The influence function as an aid in outlier detection in discriminant analysis. *Journal of the Royal Statistical Society Series B*, 27(3):251–258.

Campbell, N. A. and Atchley, W. R. (1981). The geometry of canonical variate analysis. *Systematic Zoology*, 30(3):268–280.

Campbell, N. A. and Kiiveri, H. T. (1988). Neighbour relations and remotely sensed data. Internal report, Division of Mathematics and Statistics, CSIRO, Western Australia.

Campbell, N. A. and Kiiveri, H. T. (1993). Canonical variate analysis with spatially–correlated data. *Australian Journal of Statistics*, 35(3):333–344.

Campbell, N. A. and Mahon, R. J. (1974). A multivariate study of variation in two species of rock crab of the genus leptographus. *Australian Journal of Zoology*, 22:417–425.

Charalambous, C. (1990). A conjugate gradient algorithm for the efficient training of artificial neural networks. Technical Report 90/06, Centre for Research into Sensory Technology, University of Western Australia.

Chauvin, Y. (1990). Generalized performance of overtrained back–propagation networks. In Almeida and Wellekens (1990), pp. 46–56.

Chen, D. S. and Jain, R. C. (1994). robust backpropagation learning algorithm for function approximation. *Neural Networks*, 5(3):467–479.

Cheng, B. and Titterington, D. M. (1994). Neural networks: A review from a statistical perspective (with discussion). *Statistical Science*, 9(1):2–54.

Clark, R., Swayze, G., Gallagher, A., Gorelick, N., and Kruse, F. (1991). Mapping with imaging spectrometer data using the complete band shape least-squares algorithm simultaneously fit to multiple spectral features from multiple materials. In *Proceedings of the Third Airborne Visible/Infrared Imaging Spectrometer (AVIRIS) Workshop*, pp. 2–3.

Clarke, B. R. (1983). Uniqueness and Fréchet differentiability of functional solutions to maximum likelihood type equations. *Annals of Statistics*, 11(4):1196–1205.

Clunies-Ross, C. W. and Riffenburgh, R. H. (1960). Geometry and linear discrimination. *Biometrika*, 47:185–189.

Cook, R. D. and Weisberg, S. (1982). *Residuals and Influence in Regression*. Monographs on Statistics and Applied Probability. Chapman and Hall, London.

Cooke, T. and Peake, M. (2002). The optimal classification using a linear discriminant for two point classes having known mean and covariance. Published electronically January 11, 2002, *Journal of Multivariate Analysis*.

Courant, D. and Hilbert, D. (1953). *Methods of Mathematical Physics*. Wiley-Interscience Publications, New York.

Cox, D. R. (1958). Two further applications of a model for binary regression. *Biometrika*, 45:562–565.

Cox, T. F. and Ferry, G. (1991). Robust logistic discrimination. *Biometrika*, 7(4):841–849.

Cox, T. F. and Pearce, K. F. (1997). A robust logistic discrimination model. *Statistics and Computing*, 7:155–161.

Cybenko, G. (1990). Complexity theory of neural networks and classification problems. In Almeida and Wellekens (1990), pp. 46–55.

Day, N. E. and Kerridge, D. F. (1967). A general maximum likelihood discriminant. *Biometrics*, 23:313–323.

de Boor, C. (1978). *A Practical Guide to Splines*, volume 27 of *Applied Mathematical Sciences*. Springer–Verlag, Berlin.

Diaconis, P. and Freedman, D. (1984). Asymptotics of graphical projection pursuit. *Annals of Statistics*, 12(3):793–815.

Dietterich, T. G. (1996). Statistical tests for comparing supervised classification learning algorithms. OR 97331, Oregan State University.

Dobson, A. J. (1990). *An Introduction to Generalized Linear Models*. Chapman and Hall, London.

Donoho, D. (2000). High-dimensional data analysis: The curses and blessings of dimensionality. Available at `www-stat.stanford.edu/~donoho/Lectures/AMS2000`.

Dunne, R. (1997). Evaluating contextural information models. Unpublished CMIS Remote Sensing and Monitoring Task Report, CSIRO, Floreat Park, Western Australia.

Dunne, R. (2000). A graphical method of selecting starting values for an MLP model. Unpublished CMIS Remote Sensing and Monitoring Task Report, CSIRO, Floreat Park, Western Australia.

Dunne, R. and Campbell, N. A. (1994). A suite of programs for fitting multi–layer perceptrons in Fortran and S–PLUS – Version 2.1. Technical Report 94/05, Murdoch University, Western Australia.

Dunne, R., Campbell, N. A., and Kiiveri, H. T. (1992). Task–based pruning. In *Proceedings of the Third Australian Conference on Neural Networks*, pp. 166–170. Sydney University Electrical Engineering Department.

Efron, B. (1975). The efficiency of logistic regression compared to normal discriminant analysis. *Journal of the American Statistical Association*, 70:892–898.

Efron, B. (1983). Estimating the error rate of a prediction rule. improvements on cross-validation. *Journal of the American Statistical Association*, 78:316–331.

Fahlman, S. E. (1989). Fast-learning variations on back-propagation: An empirical study. In Touretzky (1989), pp. 38–51.

Fahlman, S. E. (1990). Readme.sonar. Text file accompanying the distribution of the sonar data set of Gorman and Sejnowski.

Fisher, R. A. (1936). The use of multiple measurements in taxonomic problems. *Annals of Eugenics*, 7:179–188.

Francis, B., Green, M., and Payne, C. (1993). *The GLIM System, Release 4 Manual*. Clarendon Press, Oxford.

Frank, I. E. and Friedman, J. H. (1993). A statistical view of some chemometrics regression tools (with discussion). *Technometrics*, 35(2):109–148.

Freund, Y. and Schapire, R. E. (1996). Experiments with a new boosting algorithm. In *Machine Learning: Proceedings of the Thirteenth International Conference*, pp. 148–156.

Friedman, J., Hastie, T., and Tibshirani, R. (1998). Additive logistic regression: a statistical view of boosting. Technical report, Department of Statistics, Stanford University.

Friedman, J. H. (1984). A variable span smoother. Tech. Rep. No. 5, Laboratory for Computational Statistics, Dept. of Statistics, Stanford Univ., California.

Friedman, J. H. (1989). Regularized discriminant analysis. *Journal of the American Statistical Association*, 84:165–175.

Friedman, J. H. (1991a). Multivariate adaptive regression splines (with discussion). *Annals of Statistics*, 19:1–141.

Friedman, J. H. (1991b). Unpublished lecture notes, CSIRO, Sydney, Australia.

Friedman, J. H. (1996). On bias, variance, 0/1-loss, and the curse-of-dimensionality. Available on `http://www-stat.stanford.edu/~jhf`.

Friedman, J. H. (1999). Stochastic gradient boosting. Available at `http://www-stat.stanford.edu/~jhf/ftp/stobst.ps`.

Friedman, J. H. (2000). Greedy function approximation: A gradient boosting machine. The 1999 IMS Reitz Lecture. Available at `http://www-stat.stanford.edu/~jhf/ftp/trebst.ps`.

Friedman, J. H. and Silverman, B. W. (1989). Flexible parsimonious smoothing and additive modeling. *Technometrics*, 31(1):3–21.

Friedman, J. H. and Stuetzle, W. (1981). Projection pursuit regression. *Journal of the American Statistical Association*, 76:817–823.

Fukumizu, K. (1996). A regularity condition of the information matrix of a multi-layer perceptron network. *Neural Networks*, 9(5):871–879.

Fukunaga, K. (1990). *Introduction to Statistical Pattern Recognition*. Academic Press, Boston, second edition. [First edition, 1972].

Funahashi, K. (1989). On the approximate realization of continuous mappings by neural networks. *Neural Networks*, 2:183–192.

Furby, S., Kiiveri, H., and Campbell, N. (1990). The analysis of high dimensional spectral curves. In *Proceedings of the Fifth Australian Remote Sensing Conference*, pp. 175–184.

Gallinari, P., Thiria, S., and Fogelman-Soulie, F. (1988). Multilayer perceptrons and data analysis. In *IEEE Annual International Conference on Neural Networks*, volume 1, pp. 391–399.

Geman, S., Bienenstock, E., and Doursar, R. (1992). Neural networks and the bias/variance dilemma. *Neural Computation*, 4(1):1–58.

Geman, S. and Geman, D. (1984). Stochastic relaxation, Gibbs distributions, and the Bayesian restoration of images. *IEEE Transactions on Pattern Analysis and Machine Intelligence*, 6(6):721–741.

Gibson, G. J. and Cowan, C. F. N. (1990). On the decision regions of multilayer perceptrons. *Proceedings of the IEEE*, 78(10):1590–1594.

Girosi, F., Jones, M., and Poggio, T. (1993). Priors, stabilizers and basis functions: from regularization to radial, tensor and additive splines. A. I. Memo No. 1430 C.B.C.L. Paper No. 75, Massachusetts Institute of Technology.

Gish, H. (1990). A probabilistic approach to the understanding and training of neural network classifiers. In *IEEE International Conference on Acoustics, Speech and Signal Processing*, pp. 1361–1364.

Golub, G. H. and van Loan, C. F. (1982). *Matrix Computations*. Johns Hopkins University Press, Baltimore.

Golub, G. H. and Wilkinson, J. H. (1966). Note on iterative refinement of least squares solution. *Numer. Math*, 9:139–148.

Goodacre, R., Kell, D. B., and Bianchi, G. (1992). Neural networks and olive oil. *Nature*, 359:594.

Goodall, C. (1983). M-estimators of location: An outline of the theory. In Hoaglin, D. C., Mosteller, F., and Tukey, J. W., editors, *Understand Robust and Exploratory Data Analysis*, pp. 339–403. Wiley, New York.

Gorman, R. and Sejnowski, T. (1988). Analysis of hidden units in a layered network trained to classify sonar targets. *Neural Networks*, 1:75–89.

Goutte, C. and Hansen, L. K. (1997). Regularization with a pruning prior. *Neural Networks*, 10(6):1053–1059.

Halmos, P. R. (1974). *Measure Theory*. Springer-Verlag, New York.

Hamamoto, Y., Uchimura, S., and Tomita, S. (1996). On the behavior of artificial neural network classifiers in high–dimensional spaces. *IEEE Transactions on Pattern Analysis and Machine Intelligence*, 18(5):571–574.

Hampel, F. (1974). The influence curve and its role in robust estimation. *Journal of the American Statistical Association*, 69:383–393.

Hampel, F. R. (1985). The breakdown points of the mean combined with some rejection rules. *Technometrics*, 27(2):95–107.

Hampel, F. R., Ronchetti, E. M., Rousseeuw, P. J., and Stahel, W. A. (1986). *Robust Statistics: The Approach Based on Influence Functions*. Wiley, New York.

Hampshire, II, J. B. and Perlmutter, B. A. (1990). Equivalence proofs for multi–layer perceptron classifiers and the Bayesian discriminant function. In Touretzky, D., Elman, A., Sejnowski, T., and Hinton, G., editors, *Proceedings of the 1990 Connectionist Models Summer School*, San Mateo, CA. Morgan Kaufmann.

Hand, D. J. (1997). *Construction and Assessment of Classification Rules.* Wiley, Chichester.

Hand, D. J. (2006). Classifier technology and the illusion of progress. *Statistical Science*, 21(1):1–14. with discussion.

Hassibi, B. and Stork, D. G. (1991). Second order derivatives for network pruning: Optimal brain surgeon. In Lippmann et al. (1991), pp. 164–171.

Hastie, T. (1994). Nonparametric discriminant analysis. Unpublished lecture notes, AT&T Bell Laboratories.

Hastie, T., Buja, A., and Tibshirani, R. (1995). Penalized discriminant analysis. *Annals of Statistics*, 23:73–102.

Hastie, T. and Tibshirani, R. (1996). Discriminant analysis by Gaussian mixtures. *Journal of the Royal Statistical Society B*, 58:158–176.

Hastie, T., Tibshirani, R., and Buja, A. (1994). Flexible discriminant analysis by optimal scoring. *Journal of the American Statistical Association*, 89:1255–1270.

Hastie, T., Tibshirani, R., and Buja, A. (1999). Flexible discriminant and mixture models. In Kay, J. and Titterington, D. M., editors, *Statistics and Neural Networks: Recent Advances at the Interface.* Oxford University Press.

Hastie, T., Tibshirani, R., and Friedman, J. (2001). *The Elements of Statistical Learning: Data Mining, Inference and Prediction.* Springer.

Hastie, T., Tibshirani, R., Leisch, F., Hornik, K., and Ripley, B. D. (2006). *mda: Mixture and flexible discriminant analysis.* S original by Trevor Hastie and Robert Tibshirani. R port by Friedrich Leisch and Kurt Hornik and Brian D. Ripley. R package version 0.3-2.

Hastie, T. J. and Pregibon, D. (1992). Generalized linear models. In Chambers, J. M. and Hastie, T. J., editors, *Statistical Models in S*, chapter 9. Wadsworth and Brooks, Pacific Grove, CA.

Haykin, S. (1999). *Neural Networks. A Comprehensive Foundation.* Prentice-Hall, Engelwood Cliffs, NJ.

Hebb, D. O. (1949). *The Organization of Behavior.* Wiley, New York.

Hestenes, M. R. and Stiefel, E. (1952). Methods of conjugate gradients for solving linear systems. *Jour. of research of the national bureau of standard*, 49(6):409–436.

Himmelblau, D. M. (1990). Introducing efficient second order effects into back propagation learning. In *International Joint Conference on Neural Networks*, volume 1, pp. 631–634, Hillsdale, NJ. Lawrence Erlbaum.

Hogg, R. V. (1979). An introduction to robust estimation. In Launer, R. L. and Wilkinson, G. N., editors, *Robustness In Statistics.* Academic Press, Boston.

Hornik, K., Sinchcombe, M., and White, H. (1989). Multilayer feedforward networks are universal approximators. *Neural Networks*, 2(5):359–365.

Huber, P. J. (1981). *Robust Statistics*. Wiley, New York.

Huber, P. J. (1985). Projection pursuit. *Annals of Statistics*, 13(2):435–475.

Hunt, B., Qi, Y., and DeKruger, D. (1992). A generalization method for back-propagation using fuzzy sets. In *Proceedings of the Third Australian Conference on Neural Networks*, pp. 12–16. Sydney University Electrical Engineering Department.

Ingrassia, S. and Morlini, I. (2005). Neural network modeling for small datasets. *Technometrics*, 47(3):297–311.

Jervis, T. T. and Fitzgerald, W. J. (1993). Optimization schemes for neural networks. Technical Report CUED/F-INFENG/TR 114, Cambridge University, Electrical Engineering Department.

Jordan, M. I. (1995). Why the logistic function? A tutorial on probabilities and neural networks. Computational Cognitive Science Technical Report 9603, Massachusetts Institute of Technology.

Kärkkäinen, T. and Heikkola, E. (2004). Robust formulations for training multilayer perceptrons. *Neural Computation*, 16:837–862.

Kiiveri, H. T. (1992). Canonical variate analysis of high–dimensional spectral data. *Technometrics*, 34(3):321–331.

Kiiveri, H. T. and Caccetta, P. (1996). Some statistical models for remotely sensed data. Unpublished report, CSIRO, Floreat Park, Western Australia.

Kiiveri, H. T. and Campbell, N. A. (1992). Allocation of remotely sensed data using Markov models for image data and pixel labels. *Australian Journal of Statistics*, 34(3):361–374.

Knerr, S., Personnaz, L., and Dreyfus, G. (1990). Single layer learning revisited: A stepwise procedure for building and training a neural network. In Fogelman-Soulié, F. and Hérault, J., editors, *Neurocomputing: Algorithms, Architectures, and Applications*, volume F 68. NATO ASI Series, Springer-Verlag, Berlin.

Knerr, S., Personnaz, L., and Dreyfus, G. (1992). Handwritten digit recognition by neural networks with single–layer training. *IEEE Transactions on Neural Networks*, 3(6):962–968.

Krzanowski, W. J. and Hand, D. J. (1997). Assessing error rate estimators: the leave-one-out method reconsidered. *Australian Journal of Statistics*,, 39:35–46.

Krzanowski, W. J. and Marriott, F. H. C. (1994). *Multivariate Analysis Part 2: Classification, Covariance Structures and Repeated Measurements*. Edward Arnold, London.

Le Cun, Y., Denker, J. S., and Solla, S. A. (1990). Optimal brain damage. In Touretzky (1990), pp. 598–605.

Lee, Y. and Lippmann, R. P. (1990). Practical characteristics of neural network and conventional classifiers on artificial speech problems. In Touretzky (1990), pp. 168–177.

Li, S. Z. (1995). *Markov Random Field Modeling in Computer Vision.* Springer-Verlag, New York.

Liano, K. (1996). Robust error measure for supervised neural network learning with outliers. *IEEE Transactions on Neural Networks,* 7(1):246–250.

Liestol, K., Andersen, P. K., and Andersen, U. (1994). Survival analysis and neural networks. *Statistics in Medicine,* 13:1189–1200.

Lippmann, R. P. (1987). An introduction to computing with neural networks. *IEEE Transactions on Acoustics, Speech, and Signal Processing,* 4(2):4–22.

Lippmann, R. P., Moody, J. E., and Touretzky, D. S., editors (1991). *Advances in Neural Information Processing Systems 3. Proceedings of the 1990 Conference.* Morgan Kaufmann, San Mateo, CA.

Liu, Y. (1994). Robust parameter estimation and model selection for neural network regression. In Cowan, J. D., Tesauro, G., and Alspector, J., editors, *Advances in Neural Information Processing Systems 6. Proceedings of the 1993 Conference,* pp. 192–199, San Francisco. Morgan Kaufmann.

Lui, H. C. (1990). Analysis of decision contour of neural network with sigmoid nonlinearity. In *International Joint Conference on Neural Networks,* volume 1, pp. 655–659, Hillsdale, NJ. Lawrence Erlbaum.

Mahon, R. (1974). *A study of rock crabs of the genus Leptographus Milne Edwards.* PhD thesis, Department of Zoology, University of Western Australia.

Mardia, F., Kent, T., and Bibby, J. (1979). *Multivariate Analysis.* Academic Press, New York.

Marquardt, D. W. (1970). Generalized inverses, ridge regression, biased linear estimation, and non–linear estimation. *Technometrics,* 12(3):591–612.

Marx, B. D. and Eilers, P. H. C. (1999). Generalized linear regression on sampled signals and curves: a p-spline approach. *Technometrics,* 41(1):1–13.

Mathieson, M. (1997). Ordered classes and incomplete examples in classification. In Mozer, M. C., M. I, J., and Petsche, T., editors, *Advances in Neural Information Processing Systems,* volume 9. MIT Press, Cambridge, MA.

McCullagh, P. and Nelder, J. A. (1989). *Generalized Linear Models.* Chapman and Hall, London.

McCulloch, W. S. and Pitts, W. (1943). A logical calculus of ideas immanent in nervous activity. *Bulletin of Mathematical Biophysics,* 5:115–133. Reprinted in Anderson and Rosenfeld (1988).

McLachlan, G. J. (1992). *Discriminant Analysis and Statistical Pattern Recognition.* Wiley, New York.

Mielniczuk, J. and Tyrcha, J. (1993). Consistency of multilayer preceptron regression estimators. *Neural Networks,* 6:1019–1022.

Minsky and Papert (1969). *Perceptrons.* MIT Press, Cambridge, MA.

Mitchie, D., Spiegelhalter, D. J., and Taylor, C. C. (1994). *Machine Learning, Neural and Statistical Classification.* Ellis Horwood. The book is now out of print and has been made available at `http://www.amsta.leeds.ac.uk/~charles/statlog`.

Møller, M. F. (1993). A scaled conjugate gradient algorithm for fast supervised learning. *Neural Networks*, 6:525–533.

Moody, J. and Utans, J. (1995). Architecture selection strategies for neural networks: application to corporate bond rating prediction. In Refenes, A.-P., editor, *Neural Networks in the Capital Markets*, pp. 277–300, Chichester. Wiley.

Moody, J. E. (1992). The *effective* number of parameters: An analysis of generalization and regularization in nonlinear learning systems. In Moody, J. E., Hanson, S. J., and Lippmann, R. P., editors, *Advances in Neural Information Processing Systems 4. Proceedings of the 1991 Conference.* Morgan Kaufmann, San Mateo, CA.

Morgan, J. N. and Sonquist, J. A. (1963). Problems in the analysis of survey data, and a proposal. *Journal of the American Statistical Association*, 58:415–434.

Morgan, N. and Bourland, H. (1991). Generalization and parameter estimation in feedforward nets: some experiments. In Lippmann et al. (1991), pp. 630–637.

Movellan, J. R. (1990). Error functions to improve noise resistance and generalization in backpropagation networks. In *International Joint Conference on Neural Networks*, volume 1, pp. 557–560, Hillsdale, NJ. Lawrence Erlbaum.

Mozer, M. C. and Smolensky, P. (1989). Skeletonization: A technique for trimming the fat from a network via relevance assessment. In Touretzky (1989), pp. 107–115.

Murata, N., Yoshizawa, S., and Amari, S. (1994). Network information criterion— determining the number of hidden units for artificial neural network models. *IEEE Transactions on Neural Networks*, 5:865–872.

Nash, J. (1990). *Compact Numerical Methods For Computers. Linear Algebra and Function Minimization.* IOP Publishing, Bristol.

Neal, R. M. (1995). *Bayesian Learning for Neural Networks.* PhD thesis, Department of Computer Science, University of Toronto.

Nelder, J. A. and Mead, R. (1965). A simplex algorithm for function minimization. *Computer Journal*, 7:308–313.

Noble, B. and Daniel, J. W. (1988). *Applied Linear Algebra.* Prentice–Hall International, Englewood Cliffs, NJ.

Nowlan, S. J. and Hinton, G. E. (1992). Simplifying neural networks by soft weight-sharing. *Neural Computation*, 4:473–493.

O'Neill, R. (1971). Algorithm AS 47 – function minimization using a simplex procedure. *Applied Statistics*, 20:338–345.

Osborne, M., Presnell, B., and Turlach, B. (1998). Knot selection for regression splines via the lasso. In Weisberg, S., editor, *Dimension Reduction, Computational Complexity, and Information*, volume 30 of *Computing Science and Statistics, Interface Foundation of North America*, pp. 44–49. Fairfax Station.

Pao, Y. (1989). *Adaptive Pattern Recognition and Neural Networks*. Addison-Wesley, Reading, MA.

Paola, J. D. and Schowergerdt, R. A. (1995). A review and analysis of backpropagation neural networks for classification of remotely–sensed multi–spectral imagery. *International Journal of Remote Sensing*, 16(16):3033–3058.

Pernía-Espinoza, A. V., Ordieres-Merá, J. B., de Pisón, F. J. M., and González-Marcos, A. (2005). Tao-robust backpropagation learning algorithm. *Neural Networks*, 18(2):191–204.

Piegorsch, W. W. and Casella, G. (1989). The early use of matrix diagonal increments in statistical problems. *SIAM Review*, 31(3):428–434.

Plaut, D. C., Nowlan, S. J., and Hinton, G. E. (1986). Experiments on learning by backpropagation. Tech. Rep. CMU-CS-86-126, Carnegie Mellon University, Pittsburg.

Poggio, T. and Girosi, F. (1990a). Networks for approximation and learning. *Proceedings of the IEEE*, 78(9):1481–1497.

Poggio, T. and Girosi, F. (1990b). Regularization algorithms for learning that are equivalent to multilayer networks. *Science*, 247:978–982.

Polak, E. (1971). *Computational Methods in Optimization*. Academic Press, New York.

Pregibon, D. (1982). Resistant fits for some commonly used logistic models with medical aplications. *Biometrics*, 38:485–498.

Press, W. H., Flannery, B. P., Teukolsky, S. A., and Vetterling, W. T. (1992). *Numerical Recipes in C*. Cambridge University Press, Cambridge, second edition.

R Development Core Team (2006). *R: A Language and Environment for Statistical Computing*. R Foundation for Statistical Computing, Vienna, Austria. ISBN 3-900051-07-0.

Ramsay, J. O. and Silverman, W. B. (1997). *Functional Data Analysis*. Springer-Verlag, New York.

Ramsay, J. O. and Silverman, W. B. (1999). *S-PLUS Functions for FDA*. Manual accompanying FDA software.

Rao, C. R. (1948). The utilization of multiple measurements in problems of biological classification (with discussion). *Journal of the Royal Statistical Society B*, 10:159–203.

Rao, C. R. (1973). *Linear Statistical Inference and Its Applications*. Wiley, New York, second edition.

Reed, R. (1993). Pruning algorithms – a survey. *IEEE Transactions on Neural Networks*, 4(5):740–747.

Reed, R., Marks II, R. J., and Oh, S. (1995). Similarities of error regularization, sigmoid gain scaling, target smoothing, and training with jitter. *IEEE Transactions on Neural Networks*, 6(3):529–538.

Reinsch, C. H. (1967). Smoothing by spline functions. *Numerische Mathematik*, 10:177–183.

Rice, J. A. and Silverman, B. W. (1991). Estimating the mean and covariance structure nonparametrically when the data are curves. *Journal of the Royal Statistical Society B*, 53(1):233–243.

Richard, M. D. and Lippmann, R. P. (1991). Neural network classifiers estimate Bayesian *a posteriori* probabilities. *Neural Computation*, 3:461–483.

Richards, J. A. (1986). *Remote Sensing Digital Image Analysis: An Introduction.* Springer-Verlag, Berlin.

Richards, J. A. and Jia, X. (2006). *Remote Sensing Digital Image Analysis: An Introduction.* Springer, Berlin, 4th edition.

Ripley, B. D. (1986). Statistics, images and pattern recognition (with discussion). *Canadian Journal of Statistics*, 14:83–111.

Ripley, B. D. (1994a). Flexible non–linear approaches to classification. In Cherkassky, V., Friedman, J. H., and Wechsler, H., editors, *From Statistics to Neural Networks: Theory and Pattern Recognition*, pp. 105–126, Berlin. ASI Proceedings, subseries F, Computer and Systems Sciences, Springer–Verlag.

Ripley, B. D. (1994b). multinom. Software for fitting multinomial models via an MLP. Available from statlib at `http://lib.stat.cmu.edu/` .

Ripley, B. D. (1994c). Neural networks and related methods for classification (with discussion). *Journal of the Royal Statistical Society, B*, 56:409–456.

Ripley, B. D. (1995). Statistical ideas for selecting network architectures. In Kappen, B. and Gielen, S., editors, *Neural Networks: Artificial Intelligence and Industrial Applications*, pp. 183–190. Springer, London.

Ripley, B. D. (1996). *Pattern Recognition and Neural Networks.* Cambridge University Press, Cambridge.

Rosenblatt, F. (1962). *Principles of Neurodynamics.* Spartan Books, Washington, DC.

Rousseeuw, P. J. and Leroy, A. M. (1987). *Robust Regression and Outlier Detection.* Wiley, New York.

Rousseeuw, P. J. and van Zomeren, B. C. (1990). Unmasking multivariate outliers and leverage points (with discussion). *Journal of the American Statistical Association*, 85:633–651.

Rowan, T. (1990). *Functional Stability Analysis of Numerical Algorithms.* PhD thesis, Department of Computer Sciences, University of Texas at Austin.

Rubenstein, D. and Hastie, T. (1997). Discriminative vs informative learning. available on http://www-stat.stanford.edu/~trevor/.

Rumelhart, D. E., Hinton, G. E., and Williams, R. J. (1986). Learning representations by back-propagating errors. *Nature*, 323:533–536. Reprinted in Anderson and Rosenfeld (1988).

Sankar, A. and Mammone, R. J. (1990a). A fast learning algorithm for tree neural networks. In *Proceedings of the 1990 Conference on Information Sciences and Systems*, pp. 638–642, Princeton, NJ.

Sankar, A. and Mammone, R. J. (1990b). Tree structured neural networks. Technical Report CAIP–TR–122, Center for Computer Aids for Industrial Productivity, Rutgers University.

Sankar, A. and Mammone, R. J. (1991a). Optimal pruning of neural tree networks for improved generalization. In *Proceedings of the International Joint Conference on Neural Networks, Seattle*, volume 2, pp. 219–224.

Sankar, A. and Mammone, R. J. (1991b). Speaker independent vowel recognition using neural tree networks. In *Proceedings of the International Joint Conference on Neural Networks, Seattle*, volume 2, pp. 809–814.

Sarle, W. S. (1994). Neural networks and statistical models. In *Proceedings of the Nineteenth Annual SAS Users Group International Conference*, pp. 1538–1550.

Sarle, W. S. (1995). Stopped training and other remedies for overfitting. In *Proceedings of the 27th Symposium on the Interface*, pp. 352–360.

Sarle, W. S. (1996). Faq2.html. The Frequently Asked Questions (FAQ) list for the comp.ai.neural-nets newsgroup, June.

Saul, L. K. and Jordan, M. I. (1996). Exploiting tractable substructures in intractable networks. In Touretzky, D. S., Mozer, M. C., and Hasselmo, M. E., editors, *Advances in Neural Information Processing Systems*, volume 8, pp. 486–492, Cambridge, MA. The MIT Press.

Scarselli, F. and Tsoi, A. C. (1998). Universal approximation using feedforward neural networks: a survey of some existing methods and some new results. *Neural Networks*, 11(1):15–37.

Schowengerdt, R. A. (1997). *Remote Sensing.* Academic Press, San Diego.

Scott, D. W. (1992). *Multivariate Density Estimation.* Wiley, New York.

Serra, R. and Zanarini, G. (1990). *Complex Systems and Cognitive Processes.* Springer Verlag, Berlin.

Sethi, I. (1990). Entropy nets: from decision trees to neural networks. *Proceedings of the IEEE*, 78(10):1605–1613.

Shanno, D. F. and Phua, K. H. (1980). Remark on "Algorithm 500: Minimization of unconstrained multivariable functions". *ACM Transactions on Mathematical Software*, 6:618–622.

Shewchuk, J. R. (1994). An introduction to the conjugate gradient method without the agonizing pain. Technical Report CMU-CS-94-125, Carnegie Mellon University, School of Computer Science.

Sietsma, J. and Dow, R. J. F. (1991). Creating artificial neural networks that generalize. *Neural Networks*, 4:67–79.

Silva, F. M. and Almeida, L. B. (1990). Acceleration techniques for the backpropagation algorithm. In Almeida and Wellekens (1990), pp. 110–119.

Silverman, B. W. (1986). *Density Estimation for Statistics and Data Analysis*. Chapman & Hall, London.

Silvey, S. D. (1975). *Statistical Inference*. Chapman and Hall, London.

Smagt, van der, P. P. (1994). Minimization methods for training feedforward neural networks. *Neural Networks*, 7(1):1–11.

Smith, P. L. (1979). Splines: As a useful and convenient statistical tool. *American Statistician*, 33(2):57–62.

Sneath, P. H. A. and Sokal, R. R. (1973). *Numerical taxonomy*. Freeman, San Francisco.

Specht, D. F. (1990a). Probabilistic neural networks. *Neural Networks*, 3:109–118.

Specht, D. F. (1990b). Probabilistic neural networks and the polynomial adaline as complementary techniques for classification. *IEEE Transactions on Neural Networks*, 1:111–121.

Specht, D. F. (1991). A general regression neural network. *IEEE Transactions on Neural Networks*, 2:568–576.

Staudte, R. G. and Sheather, S. J. (1990). *Robust Estimation and Testing*. Wiley, New York.

Stewart, G. W. (1987). Collinearity and least squares regression. *Statistical Science*, 2:68–100.

Strang, G. (1988). *Linear Algebra and its Applications*. Harcourt Brace Jovanovich, San Diego, third edition.

Su, J. Q. and Liu, J. S. (1993). Linear combinations of multiple diagnostic markers. *Journal of the American Statistical Association*, 88(424):1350–1355.

Tibshirani, R. (1996). Regression shrinkage and selection via the LASSO. *Journal of the Royal Statistical Society B*, 58(1):267–288.

Titterington, D. M., Murray, G. D., Murray, L. S., Spiegelhalter, D. J., Skene, A. M., Habbema, J. D. F., and Gelpka, G. J. (1981). Comparison of discrimination techniques applied to a complex data set of head injured patients (with discussion). *Journal of the Royal Statistical Society A*, 144:145–174.

Touretzky, D., Hinton, G., and Sejnowski, T., editors (1988). *Proceedings of the 1988 Connectionist Models Summer School, Pittsberg.* Morgan Kaufmann, San Mateo, CA.

Touretzky, D. S., editor (1989). *Advances in Neural Information Processing Systems. Proceedings of the 1988 Conference.* Morgan Kaufmann, San Mateo, CA.

Touretzky, D. S., editor (1990). *Advances in Neural Information Processing Systems 2. Proceedings of the 1989 Conference.* Morgan Kaufmann, San Mateo, CA.

Turing, A. M. (1948). Rounding-off errors in matrix processes. *Quart. J. Mech*, pp. 287–308.

Utans, J. and Moody, J. (1991). Selecting neural network architectures via the prediction risk: application to corporate bond rating prediction. In *Proc. of the First Int. Conf on AI Applications on Wall Street*, Los Alamos, CA. IEEE Computer Society.

Vapnik, V. N. (1995). *The Nature of Statistical Learning Theory.* Springer, New York.

Venables, W. N. and Ripley, B. D. (1994). *Modern Applied Statistics with S-Plus.* Springer, New York, first edition.

Venables, W. N. and Ripley, B. D. (1999). *Modern Applied Statistics with S-Plus.* Springer, New York, third edition.

Venables, W. N. and Ripley, B. D. (2000). *S Programming.* Springer, New York.

Venables, W. N. and Ripley, B. D. (2002). *Modern Applied Statistics with S.* Springer, New York, fourth edition.

Wahba, G. (1990). *Spline Models for Observational Data*, volume 59 of *Regional Conference in Applied Mathematics.* S.I.A.M.

Wand, M. P. (1999). A comparison of regression spline smoothing procedures. Unpublished manuscript available at http://www.biostat.harvard.edu/~mwand/papers.html.

Webb, A. and Lowe, D. (1990). The optimized internal representation of multilayer classifier networks performs nonlinear discriminant analysis. *Neural Networks*, 3:367–375.

Webb, A., Lowe, D., and Bedworth, M. (1988). A comparison of nonlinear optimization strategies for feed–forward adaptive layered networks. Technical Report 4157, Royal Signal and Radar Establishment.

Weigend, A. S. (1994). On overfitting and the effective number of hidden units. In Mozer, M. C., editor, *Proceedings of the 1993 Connectionist Models Summer School*, pp. 335–342, Hillsdale, NJ. Lawrence Erlbaum.

Weigend, A. S., Rumelhart, D. E., and Huberman, B. A. (1991). Generalization by weight-elimination with application to forecasting. In Lippmann et al. (1991), pp. 875–882.

Werbos, P. (1974). *Beyond Regression*. PhD thesis, Harvard University.

White, H. (1989). Learning in artificial neural networks: A statistical perspective. *Neural Computation*, 1:425–464.

Widrow, B. and Lehr, M. A. (1990). 30 years of adaptive networks: perceptron, madaline and backpropagation. *Proceedings of the IEEE*, 78(9):1415–1442.

Wilkinson, G. G., Fierens, F., and Kanellopoulos, I. (1995). Integration of neural and statistical approaches in spatial data classification. *Geographical Systems*, 2:1–20.

Williams, P. (1995). Bayesian regularization and pruning using a Laplacian prior. *Neural Computation*, 7:117–143.

Wilson, J. (1992). A comparison of procedures for classifying remotely-sensed data using simulated data sets incorporating autocorrelations between spectral responses. *Int. J. Remote Sensing*, 13(14):2701–2725.

APPENDIX A

FUNCTION MINIMIZATION

A.1 INTRODUCTION

Learning consists of minimizing the discrepancies between the observed and fitted values. To implement this in a neural network, a penalty function $\rho = \rho(z^*, t)$ is imposed, such that the function has a minimum value when the output values equal the target values. We then minimize ρ over the unknown parameters.

Minimization techniques can be divided into two types: local, which find a local minimum; and global[1], which attempt to find the global minimum. Generally, global minimization is more desirable but it is also more computationally demanding and, due to the complexity of the models and the potential size of the data sets, it is often impractical. The most widely used methods for neural networks are a class of local minimization schemes based on gradient information.minimization techniques

There is now a vast literature on function minimization as applied to neural networks. Battiti (1992) gives a survey of the main characteristics of the different methods and their mutual relations, sub–titled "between steepest descent and Newton's method". The paper makes the point that steepest descent uses first–order information and is computationally cheap per iteration, whereas Newton's method uses the full second–order information and is computationally expensive per iter-

---

[1] Global techniques include genetic algorithms; simulated annealing; and perhaps Bayesian learning which integrates over the parameter space rather than finding the minima.

A Statistical Approach to Neural Networks for Pattern Recognition by Robert A. Dunne   **245**
Copyright © 2007 John Wiley & Sons, Inc.

ation. Many minimization schemes attempt to gain the benefits of second–order information without incurring the full computational costs. Recent surveys of the area can be found in Bishop (1995a), Ripley (1996), and Smagt (1994).

Barnard and Cole (1989) give details of a conjugate gradient implementation and Charalambous (1990) gives an interesting variant on conjugate gradient, using sub–vectors of the weights rather than the whole weight vector, that may be feasible for parallel MLP architectures. Møller (1993) recommends a version of conjugate gradients called "scaled conjugate gradients" and Fahlman (1989) gives an algorithm called "quickprop". Both of these appear widely in the neural network literature. Jervis and Fitzgerald (1993) conclude that scaled conjugate gradient and quickprop are both expedient minimization schemes. Himmelblau (1990) treats the problem of calculating the MLP Hessian matrix in parallel.

In the case where the MLP is used for function approximation, rather than classification, the lack of a non–linear activation function at the output layer means that, for given inputs and Ω matrix, determining the Υ matrix is a linear problem. This means that a linear approach (Gauss–Markov) can be used to estimate Υ and then a non–linear one to estimate Ω, at each learning iteration. This appears to have been considered independently by Webb et al. (1988) and Barton (1991).

Fahlman (1989) gives a discussion of the problem of comparing minimization algorithms. The paper raises the issues of:

- benchmark problems;

- stopping criteria (which can make a large difference to the reported learning times);

- reporting of training times, including best, worst, average and the question of non–converging trials.

There is also a vast numerical analysis literature on function minimization, see Nash (1990), Adby and Dempster (1974), Rowan (1990), Nelder and Mead (1965), O'Neill (1971), Shanno and Phua (1980) and Press et al. (1992).

A.2 BACK–PROPAGATION

Back–propagation is a very simple idea and is no more than the insight that if the activation function is differentiable, then the objective function can be minimized by well–known techniques. It has appeared several times in the literature, common early references being Werbos (1974) and Rumelhart et al. (1986).

Back–propagation is just steepest descent minimization, which is a well–understood process. For the purposes of this discussion, we consider ρ as a function of the vector $\omega = \{\text{vec}(\Upsilon^\tau)^\tau, \text{vec}(\Omega^\tau)^\tau\}^\tau$; then, if the activation functions, f_q and f_h are differentiable, ρ will be differentiable with respect to ω. Now, if the function ρ is expanded as far as the second term of its Taylor series, we have $\rho(\omega + \Delta\omega) = \rho(\omega) + \dfrac{\partial \rho}{\partial \omega}\Delta\omega$. Then, noting that

$$\Delta\rho = \frac{\partial \rho}{\partial \omega}\Delta\omega = |\frac{\partial \rho}{\partial \omega}||\Delta\omega|\cos\theta$$

has a maximum at $\theta = 0$ and a minimum at $\theta = \pi$, we can see that to minimize ρ, $\Delta\omega$ should be in the direction of the negative gradient $-\partial\rho/\partial\omega$.

We argue that the important insight of back–propagation is not the use of steepest descent. It is the observation that in order to use a minimization technique that requires derivatives in a parallel implementation of an MLP, certain information must be available at each unit in order to evaluate the derivative. An examination of the derivatives $\partial\rho/\partial w_{ph}$ for $\omega \in \Omega$, in Section 2.2 (p. 12), shows that the term

$$\sum_{q=1}^{Q}(z_q^* - t_q)\frac{\partial f(z_q)}{\partial z_q}u_{q,h+1}$$

appears in each derivative, and an examination of the structure of Figure 2.1 shows that the connections are in place for this information to "back–propagate" through the network along the established connections. Hence, by a cyclical process in which information first flows one way through the network and then flows the other way, the adjustments Δw can be made with information that is locally available to each unit. It is this localization of information that allows the process to be implemented in parallel.

The standard "back–propagation" algorithm for each step is

$$\Delta\omega^{(m)} \leftarrow \eta\frac{\partial\rho}{\partial\omega^{(m)}} + \alpha\Delta\omega^{(m-1)}, \tag{A.1}$$

where m is the step counter, and then

$$\omega^{(m+1)} = \omega^{(m)} - \Delta\omega^{(m)},$$

Note that (A.1) includes a "momentum" term involving the previous step, and that the step sizes, α and η, are fixed. The selection of the parameters α and η has been described as "a black art" and the best advice that is available is to make them small, as large values may prevent convergence, but not "too" small or the number of iterations needed before the algorithm converges may be unacceptable.

Due to the slow rate of convergence of the back–propagation algorithm, we have considered more sophisticated numerical tools for function minimization.

A.3 NEWTON-RAPHSON

The Newton-Raphson technique relies on the function ρ being twice differentiable. If we expand ρ near to a minimum point, we have

$$\rho(\omega + \Delta\omega) = \rho_{\min} = \rho(\omega) + \frac{\partial\rho}{\partial\omega}^T\Delta\omega + \frac{1}{2}\Delta\omega H\Delta\omega + \dots,$$

where H is the Hessian matrix of second derivatives

$$\langle H\rangle_{r_1 r_2} = \frac{\partial^2\rho(\omega)}{\partial\omega_{r_1}\partial\omega_{r_2}}\bigg|_{\omega}.$$

For convenience, we write $\frac{\partial\rho}{\partial\omega} = g(\omega)$. Taking the first two terms and differentiating with respect to x, and setting the result to 0, yields $g = -H\Delta\omega$ or $\Delta\omega = -H^{-1}g$. This is the basis for the Newton–Raphson algorithm

$$\omega^{(m+1)} = \omega^{(m)} - \alpha^{(m)}(H^{(m)})^{-1}g^{(m)}. \tag{A.2}$$

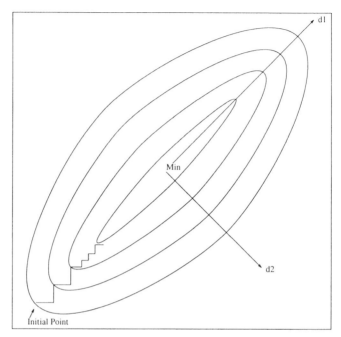

Figure A.1   A simple gradient descent method will take a zig–zag path from x_0 to the minimum and, if it finds the minimum in each search direction, the successive search directions will be orthogonal to each other. This is the reason why the momentum term in the "back–propagation" algorithm speeds up the descent. In a region like the one illustrated, since all of the steps are moving in the same general direction, a combination of search directions should minimize the zig–zag and capture the general trend towards the minimum.

The term $\alpha^{(m)}$ is found by a line search in the search direction $(H^{(m)})^{-1}g^{(m)}$ and is necessary if the function is not in fact quadratic. If ρ is quadratic, then one step will give the minimum point.

As Adby and Dempster (1974, p. 94) point out, the reason that gradient descent is slow is that the direction of steepest descent from a point and the direction of the minimum may be nearly orthogonal. Multiplying the gradient by H^{-1} changes the coordinate system to one where the function contours are spherical and the two directions coincide.

As we can see from Section 2.1, calculating the second derivatives may impose a considerable computational burden if the calculation must be repeated for each iteration[2]. Because of this, the Newton–Raphson algorithm is principally important as the theoretical under–pinning of a number of other methods, such as quasi-Newton methods, that attempt to approximate the inverse Hessian rather than calculate it at each step.

---

[2]For an $n \times n$ matrix, calculating an inverse requires $O(n^3)$ operations. Hence it is desirable to avoid doing it too often in a computation.

If we take the Taylor series of ρ about $\omega_{\min}$, we see that

$$\rho(\omega) = \rho_{\min} - g_{\min}^{\tau} \Delta\omega + \frac{1}{2} \Delta\omega \, H_{\min} \Delta\omega$$

which, as $g_{\min} = 0$, implies that

$$\Delta\omega \, H_{\min} \Delta\omega = 2\{\rho(\omega) - \rho(\omega_{\min})\} > 0$$

and therefore $H_{\min}$ is positive definite. This means that the eigenvalues of $H_{\min}$ are all strictly positive and so the determinant is greater than 0 and $H_{\min}$ is invertible. However, in general this may not be true for points away from the minimum and in the case of ρ, it is known to be untrue. The function ρ can have large flat areas, due to the fact that when one of the component weights is large it dominates the output, and changes in the other weights will not affect the value of ρ to a significant degree.

A.4   THE METHOD OF SCORING

The "method of scoring" is related to Newton-Raphson, and is widely used in the statistical literature. It is applicable in the case where ρ is a log likelihood for a distribution f and we have a sample of size N, giving a matrix X of explanatory variables ($N \times p$) and a response variable Y.

The method involves replacing H by its expected value $\mathbb{E}(H)$, so that (A.2) becomes

$$\omega^{(m+1)} = \omega^{(m)} - \{\mathbb{E}(H)^{(m)}\}^{-1} g^{(m)}$$

and thus

$$\mathbb{E}(H)^{(m)} \omega^{(m+1)} = \mathbb{E}(H)^{(m)} \omega^{(m)} - g^{(m)} \qquad (A.3)$$

(dropping the step–length term). $\mathbb{E}(H)$ is Fisher's information matrix (Silvey, 1975), and it can be shown that

$$\mathbb{E}(H) = \mathbb{E}(gg^{\tau}).$$

This is taken to mean that at the mth step, the mth estimate of $\mathbb{E}(H)$ is given by $\mathbb{E}(gg^{\tau})$.

Furthermore, in the case where f is from the exponential family, writing $\eta = X\omega$ as the linear predictor, it can be shown (Dobson, 1990) that

$$\mathbb{E}(H) = X^{\tau} W X,$$

where W is a diagonal matrix with entries

$$\langle W \rangle_{nn} = \frac{1}{\mathbb{V}(y_n)} \left(\frac{\partial \mathbb{E}(y_n)}{\partial \eta_n} \right)^2.$$

If we had a linear equation $y = X\omega$ we can solve the normal equations $X^{\tau} X \omega = X^{\tau} y$ to get ω. In this case, writing $\mu = \mathbb{E}(Y)$ and z as the *adjusted dependent variable*

$$z^{(m)} = \hat{\mu}^{(m)} + (y - \hat{\mu}^{(m)}) \left(\frac{\partial \eta}{\partial \mu} \right)^{(m)},$$

we derive an expression of the form

$$X^T W X \omega = X^T W z \tag{A.4}$$

(McCullagh and Nelder, 1989; Dobson, 1990), which must be solved iteratively for ω as both W and z may depend on ω. From (A.4) the method acquires the name "iteratively re–weighted least squares" and is widely used for fitting generalized linear models.

A.5   QUASI-NEWTON

Quasi–Newton methods attempt to approximate the matrix H^{-1} using a matrix A such that

$$\lim_{m \to \infty} A^{(m)} = H^{-1}.$$

A is determined using the step–size information and the gradient information.

The code available in Dunne and Campbell (1994) uses the "Broyden–Fletcher–Goldfarb–Shanno" algorithm. Let the search direction be $d = Ag(\omega)$ and $\Delta\omega = \alpha d$, where α is determined by a line search. Then write $\Delta g = g(\omega + \Delta\omega) - g(\omega)$ and

$$A^{(m+1)} = A^{(m)} - \frac{gg^T}{g^T d} + \frac{\Delta g (\Delta g)^T}{\alpha (\Delta g)^T d}, \tag{A.5}$$

where each of g, d and Δg are evaluated at the mth iteration. This method requires the storage of an $R \times R$ matrix; however, it has the advantage that it is well–understood and there is a substantial literature on such questions as numerical accuracy and the handling of round–off errors. See Polak (1971) for details and for a proof that (A.5) converges to H^{-1} in R steps if the function is a quadratic form. See also Press et al. (1992) for a discussion of various implementation issues.

A.6   CONJUGATE GRADIENTS

The conjugate gradient method (Hestenes and Stiefel, 1952) also attempts to overcome the problem that the direction of steepest descent and the direction to a local minimum may not be the same. Consider a quadratic function of two variables with level curves as shown in Figure A.1. A simple gradient descent method will take a zig–zag path from ω_0 to the minimum and, if it finds the minimum in each search direction, the successive search directions will be orthogonal to each other. In a region such as the one illustrated, as all of the steps are moving in the same general direction, a combination of search directions should minimize the zig–zag and capture the general trend towards the minimum. This is the reason why the momentum term in the "back–propagation" algorithm speeds the descent.

The conjugate gradient method, instead of using a fixed fraction of the previous search direction in determining the correct step, calculates the optimal, orthogonal search directions. To see how this is done, consider again the Taylor expansion of ρ,

$$\rho(\omega + \Delta\omega) \approx \rho(\omega) + g(\omega)\Delta\omega + \frac{1}{2}\Delta\omega H \Delta\omega.$$

Now, at a local minimum, the first derivative will be zero so the behavior of the function will be approximately described by the quadratic form $\Delta \omega H \Delta \omega$.

As we have already seen, H will be positive definite (assuming it is of rank R) near to a minimum point. This implies that the eigenvalues of H are non–negative and we can interpret them, and the associated eigenvectors, in a geometric way. The quadratic form $\Delta \omega H \Delta \omega$ describes an ellipse in R–dimensions and the axes of the ellipse point towards the eigenvectors of H. The eigenvalues define the width of the ellipse along its axes.

Returning to Figure A.1, for the directions labeled as d_1 and d_2, the minimum can be reached in two steps by minimizing the function along each of these directions (assuming an exact line search algorithm is available). Similarly in R–dimensions, if ω_0 is the current estimate of the minimum and we have the vectors $d_1, ..., d_R$, then the minimum can be written as $\omega_{\min} = \omega_0 + \sum_{r=1}^{R} \alpha_r d_r$, and the values of the α_r can be found by a linear search in each of the directions d_r.

As H is a real symmetric matrix, it is normal $(H^T H = H H^T)$ and so there exists an orthogonal matrix U such that

$$U^T H U = \Lambda,$$

where Λ is a diagonal matrix with the eigenvalues of H on the main diagonal. The columns of U are the associated eigenvectors. Hence if we can find the eigenvectors of H, we will have a set of R orthogonal directions along which we can minimize the function. These directions are termed conjugate directions with respect to the matrix H.

Having shown that a set of conjugate directions exists, the problem is how to determine such a set from the available information, in particular, from the gradients alone without determining the Hessian matrix.

Say that we have a set of R gradient vectors $\{g_r\}_{r=1}^{R}$ that span the space. The algorithm is then as follows:

$$\text{let } d_1 = -g_1,$$

and

$$d_r = -g_{r-1} + \sum_{j=1}^{r-1} \frac{g_r^T H d_j}{d_j^T H d_j} d_{r-1} \tag{A.6}$$

$$\text{for } r = 2, \ldots, R.$$

This is the Gram–Schmidt algorithm for finding R orthogonal directions with respect to the inner product $\langle a, b \rangle = a^T H b$ (Strang, 1988; Noble and Daniel, 1988).

Note that, for a quadratic function,

$$g_r^T g_j = 0 \text{ for } j = 1, ..., r - 1, \tag{A.7}$$

and that

$$H d_j = g_{j+1} - g_j, \tag{A.8}$$

$$d_j^T g_{j+1} = 0, \tag{A.9}$$

and

$$d_j^T g_j = -g_j^T g_j.$$ (A.10)

Then, using (A.7), (A.6) can be simplified to

$$d_r = -g_{r-1} + \sum_{j=1}^{r-1} \frac{g_r^T (g_{j+1} - g_j)}{d_j^T (g_{j+1} - g_j)} d_{r-1}$$

and using (A.8), this can be written as

$$d_r = -g_{r-1} + \frac{g_r^T (g_r - g_{r-1})}{d_{r-1}^T (g_r - g_{r-1})} d_{r-1},$$ (A.11)

which is the Hestenes–Stiefel updating formula. Using (A.9) and (A.10), we can write

$$d_r = -g_{r-1} + \frac{g_r^T (g_r - g_{r-1})}{g_{r-1}^T g_{r-1}} d_{r-1},$$ (A.12)

which is the Polak–Ribière formula, and

$$d_r = -g_{r-1} + \frac{g_r^T g_r}{g_{r-1}^T g_{r-1}} d_{r-1},$$ (A.13)

which is the Fletcher-Reeves updating formula.

Hence the conjugate gradient algorithm only requires the storage of two vectors, a search direction and a gradient, at each iteration. It is common practice to reset the search direction after a cycle of R iterations. Smagt (1994) discusses two such strategies, the gradient restart (set the search direction to $-g$) and the Beale restart algorithm. (See Shewchuk (1994) for a detailed discussion.)

A.7   SCALED CONJUGATE GRADIENTS

This algorithm, due to Møller (1993), can be viewed as a variant of the quasi–Newton algorithms. However, instead of estimating H^{-1}, we estimate H as the difference in first derivatives at successive points.

If d is the current step direction, then

$$\hat{H}d \approx \frac{g(\omega + \sigma d) - g(\omega)}{\sigma}, \qquad 0 < \sigma << 1$$

and then the inner product $d^T \hat{H} d$ is formed as for (A.6). However, it is often the case that $d^T \hat{H} d < 0$, indicating that the Hessian is not positive–definite at that point, or that the approximation to the Hessian is poor.

A line search could perhaps be used to determine a satisfactory step size at this point, but this would raise the computational complexity per iteration. The approach adopted is based on the Levenberg and Marquart algorithm (Adby and Dempster 1974) which uses

$$\hat{H}d \approx \frac{g(\omega + \sigma d) - g(\omega)}{\sigma} + \alpha d$$

and α is chosen to ensure that $d^T y \hat{H} d > 0$. Møller (1993) gives an iterative algorithm for determining α. Ripley (1996, Section A.5) comments that the algorithm uses the out–dated Hestenes–Stiefel formula (A.11) and that Møller's comparisons with quasi–Newton are invalid. Bishop (1995a, Section 7.9) comments that the algorithm is sometimes significantly faster than conjugate gradients.

A.8   VARIANTS ON VANILLA "BACK–PROPAGATION"

Due to the slow rate of convergence exhibited by the vanilla "back–propagation" algorithm, a number of authors have proposed various improvements.

Fahlman (1989), in particular, has proposed a number of refinements exploiting the functional form of the activation function. As the back–propagated error term contains the derivative of the sigmoid function, which varies from 0 to 0.25, even if the error is large for a particular pattern and a particular unit, the corresponding error term may be close to 0. Hence the unit may be frozen in an incorrect position rather than moving to a better position. (As the derivative will not be exactly 0, the unit may move unacceptably slowly, or numerical underflow may set the small derivative to 0.) Fahlman's solution is to add a small constant such as 0.1 to the derivative to prevent the unit sticking.

In a similar vein, Fahlman considers modifying the penalty function to $\rho = \mathrm{arctanh}(z_q^* - t_q)$ so that the error may be arbitrarily large. This should increase the learning rate by forcing the network to move to correct output units that are far from their desired target.

It may be the case that these approaches are better suited to noise–free problems such as the N–M–N encoder and XOR problems that Fahlman (1989) investigates. For a problem with noisy data, particularly with outliers, these methods may encourage over–fitting.

The other approaches to improving vanilla back–propagation are all variants on introducing a second–order effect. For example, the "quickprop" algorithm (Fahlman, 1989) assumes that the surface can be locally approximated by a parabola and then, using $g^{(m)}$ and $g^{(m-1)}$ takes a step to the minimum of the parabola. The step is given by

$$\Delta\omega^{(m)} = \frac{-g^{(m)}}{g^{(m-1)} - g^{(m)}}\Delta\omega^{(m-1)}.$$

A limiting parameter must be set so that if the steps are in the same direction for a number of iterations and the gradient is monotonically increasing, the step size does not grow too large.

Becker and le Cun (1989) propose a "pseudo–Newton" scheme where

$$\Delta\omega_r = \frac{-\dfrac{\partial\rho}{\partial\omega_r}}{\dfrac{\partial^2\rho}{\partial\omega_r^2} + \epsilon}$$

i.e. only the diagonal terms of the Hessian are used. The ϵ term serves the same purpose as the practice of adding a small quantity to the diagonal values of $\hat{H}$ to improve the conditioning.

Silva and Almeida (1990) propose a scheme whereby the size of the step is reduced if the sign of the gradient changes at successive iterations and is increased otherwise. This is intended to stop the algorithm stepping back and forth across the bottom of narrow gullies, something that traps gradient descent and slows convergence considerably.

All of these approaches are claimed by their authors to improve vanilla back–propagation by considerable margins.

A.9 LINE SEARCH

None of these methods is guaranteed to move in a downhill direction at any particular step. All that they do is select a direction in which the function decreases; however, too large a step in that direction may result in the function increasing. The conjugate gradient and quasi–Newton methods are generally implemented with a line search algorithm to ensure that a downhill step is taken at each iteration, while scaled conjugate gradient uses a different procedure to estimate a correct step size.

It is noted in the literature that conjugate gradient requires a more accurate line search than quasi–Newton (Ripley, 1996, Section A.5). Consider Figure A.1, which shows the level curves for a quadratic. The quasi–Newton algorithm will go the the minimum in one step, whereas the conjugate gradient algorithm requires two steps, and the minimum must be found along each search direction in order for the local minimum to be found. Hence in this setting the conjugate gradient algorithm is more dependent on a line search than quasi–Newton, and the dependence is compounded in higher dimensions.

A.10 THE SIMPLEX ALGORITHM

The simplex algorithm was introduced by Nelder and Mead (1965). See also Rowan (1990), O'Neill (1971) and Press et al. (1992) for implementations[3].

Consider an R–dimensional space and visualize a simplex moving through the space, seeking the point with the lowest value. Specializing this to three–space and the simplex $ABCD$ (Figure A.2), the steps of the algorithm are, briefly:

- evaluate f at A, B, C and D – we assume that f is a maximum at A among the points;

- reflect A through E, the centroid of BCD, to F;

- if unsuccessful ie. if $f(F) > \{f(B), f(C), f(D)\}$, then contract to G or H depending on which of $\{f(F), f(A)\}$ is higher ;

- in the case that $f(G) > f(A)$ or $f(H) > f(F)$, contract the simplex around one of $\{B, C, D\}$ depending on which is the lowest;

[3]Note that this is not same as the identically named simplex method for linear programming (Adby and Dempster, 1974).

- if successful, extend to I and keep $f(F)$ or $f(I)$, whichever is the lower,

- replace A by the new point and continue.

The step lengths suggested in Nelder and Mead (1965) are

$$\frac{EF}{AE} = 1 \quad \frac{AE}{EI} = 2 \quad \frac{AE}{EH} = 0.5.$$

A.11   IMPLEMENTATION

The authors' software for fitting MLP models (Dunne and Campbell, 1994) implements the two–layer network described in Section 2.1 (p. 9), and gives a choice of the seven minimization routines listed in Table A.1. Little attempt has been made to optimize the code; however, this should not substantially affect the comparison of the various methods.

Steepest descent, at least in the usual implementation in the neural network literature, does not include a line search. The common practice is to keep the step size small enough so that an up–hill step is not too disastrous for the algorithm. For the purposes of comparison, it has been implemented here both with and without a line search.

A.12   EXAMPLES

The seven algorithms were run on each of three examples. The MLP models used a sum of squares penalty function and no skip– or intra–layer connections. The target values were encoded with a *one of q* encoding.

example 1 is the simple case of two linearly–separable classes in a two–dimensional feature space. It is Experiment 2 from Section 9.3.2 (p. 147) and is shown in Figure 9.4 (p. 147).

example 2 is similar to example 1 except that the classes are no longer linearly separable. It is Experiment 3 from Section 9.3.3 (p. 149) and is shown in Figure 9.8 (p. 151).

example 3 is the remotely sensed data set used in Section 11.3.2 (p. 190). It consists of spectral data (sampled at 169 wavelengths) for two ground–cover classes: a field sown with potatoes; and one with alfalfa. Figure 11.5 (p. 191) shows that the spectra are not readily separable. The data were scaled to the interval $[0, 1]$ so that the large values (many thousand) of the spectral data did not cause units to become "stuck", the MLP was of size 23.1.1 and used the basis–fitting algorithm described in Chapter 11[4], p. 179.

The comparisons were based on a single trial from the same starting values. Each algorithm was allowed to run for 2000 function evaluations, whether these evaluations involved calculating the derivatives or not, and the value of ρ was reported every 10th function evaluation. The stopping tolerance was set to 0.

---

[4]It does not make sense in this example to follow the widely recommended practice of scaling each variable separately to $[0, 1]$, as the spectral structure would then be lost.

The results can be seen in Figures A.3, A.4 and A.5. It can be seen that simplex does 2000 iterations faster than the other algorithms, since it does not involve derivative calculations. In Figures A.3 and A.4, it can also be seen that all the algorithms, except steepest descent, achieve low values of ρ much sooner than the simplex algorithm. Figure A.5 is more complex. Algorithm 6 (the Polak–Ribière algorithm with Beale restarts) and algorithm 1 failed completely (they are outside the scale of the graph). Algorithm 3 (scaled conjugate gradients) also did quite badly.

A.13   DISCUSSION AND CONCLUSION

While these three experiments can hardly be called an exhaustive survey of the area, they do allow a number of observations:

- steepest descent is very slow to converge and is probably not a feasible algorithm in practice; however, the addition of a line search improves the performance of the algorithm considerably;

- simplex does remarkably well for an algorithm that uses neither derivatives nor a line search[5]. It is not as fast as the conjugate gradient and quasi–Newton algorithms in some cases, but it does not fail in any of the three experiments. As it doesn't need derivatives (and is still reasonably fast), it has been found useful in developing and debugging new network architectures;

- it is reported (Smagt, 1994) that steepest descent (algorithm 1) is less sensitive to local minima than many more sophisticated algorithms (algorithms 2–6), and this also appears to be the case with the simplex algorithm (see example 3).

An examination of the neural network literature on function minimization reveals a definite trend. Earlier works considered a range of *ad hoc* methods and implementations of more standard methods taken from the numerical analysis literature. In addition, there was a blurring of the distinction between the model being fitted and the fitting procedure (e.g., see Benediktsson et al. (1995) where reference is made to a "conjugate gradient backpropagation network"). In actual fact, the model and the fitting algorithm are totally different questions.

The trend in later works has been to consider the question of fitting the model as one rightly belonging in the domain of numerical analysis (there is not much that is special about a function made up of sums of sigmoids). The best advice that can be offered is to adopt the current best unconstrained general function minimization routine from the numerical analysis literature. This appears to be the BFGS quasi–Newton method, or conjugate gradients with the Polak–Ribière algorithm with Beale restarts, and a good line–search algorithm.

However, in the case where derivatives are not available, or less sensitivity to local minima is required, then the simplex algorithm has been found to be very useful.

---

[5]The algorithm itself could be said to incorporate a line search; however, compared to the line search used in the other algorithms, it is very rudimentary.

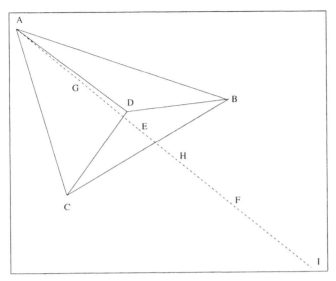

Figure A.2   A schematic representation of the action of the simplex algorithm.   See Section A.10 for details.

Table A.1   The seven minimization schemes implemented in the software of Dunne and Campbell (1994). See Smagt (1994) for details of 1 to 6 and Nelder and Mead (1965), Rowan (1990), O'Neill (1971) or Press et al. (1992) for details of 7.

1	steepest descent
2	steepest descent with a line search
3	conjugate gradients (Fletcher–Reeves algorithm with gradient restarts)
4	scaled conjugate gradients Møller (1993)
5	quasi–Newton (BFGS algorithm)
6	conjugate gradients (Polak–Ribière algorithm with Beale restarts)
7	simplex algorithm

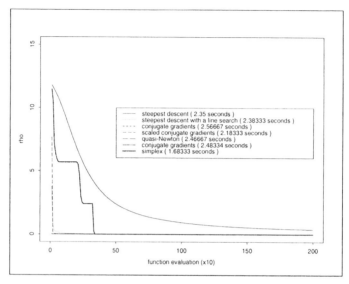

Figure A.3   The results from example 1. The value of ρ is plotted against the number of function evaluations (times 10) for the seven algorithms.

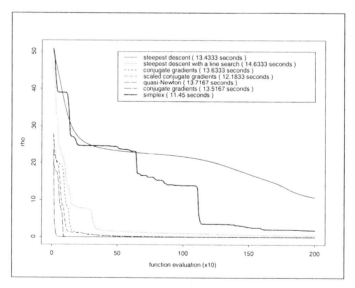

Figure A.4   The results from example 2. The value of ρ is plotted against the number of function evaluations (times 10) for the seven algorithms.

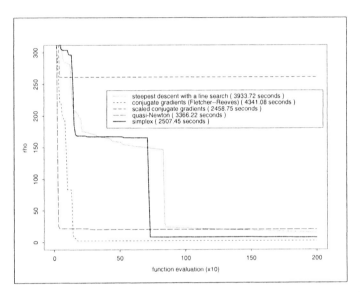

Figure A.5   The results from example 3. The value of ρ is plotted against the number of function evaluations (times 10) for five of the the seven algorithms. The curves for steepest descent and conjugate gradients (Polak–Ribière algorithm with Beale restarts) are outside the range of the plot.

APPENDIX B

MAXIMUM VALUES OF THE INFLUENCE CURVE

Restricting our attention to the case of two classes, so that the decision boundary is the surface such that $\mathrm{mlp}(x) = 0.5$, we show that the maximum value of $IC(x, \omega)$ occurs when the point x is misclassified. We show this explicitly only for the weight v_0, in an MLP of size 1.1.1, so that the Υ matrix consists of two weights v_0 and v_1, and the Ω matrix also consists of two weights, ω_0 and ω_1. The fitted model is then

$$\mathrm{mlp}(x) = \cfrac{1}{1 + \exp\left[-\left(v_0 + \cfrac{v_1}{1 + \exp\{-(\omega_0 + \omega_1 x)\}}\right)\right]}.$$

However, a similar argument is possible for each of the weights in the MLP. In addition, the argument may be extended as follows: for a weight $v_i \in \Upsilon$ where h or q are greater than 1, the derivative $\dfrac{\partial \psi}{\partial v_i}$ will not depend on the weights v_j for $j \neq i$. Hence we only need to consider a subset of the MLP, corresponding to an MLP of size p.1.1. Consider $\omega_i \in \Omega$ where p, h are greater than 1. If h is greater than 1, we only need to consider the relevant hidden layer unit to which ω_i is connected. If $p > 1$, then by considering $x = \kappa \in \ell \perp \omega^T x = 0$, we can reduce the problem to the one–dimensional case.

Now we take the limits of the product $\omega_1 x$ so that we do not have to worry about the sign of ω_1

$$\lim_{\omega_1 x \to \infty} \text{mlp}(x) = \frac{1}{1 + \exp(-v_0 - v_1)}, \tag{B.1}$$

and

$$\lim_{\omega_1 x \to -\infty} \text{mlp}(x) = \frac{1}{1 + \exp(-v_0)}. \tag{B.2}$$

We assume that we have fitted the MLP model to a two–class data set and that the objective function and the misclassification error have been successfully minimized for the model. Of the two limits, (B.1) and (B.2), we must have one that approaches a target value that is "near" to 1, and the other must approach a target value that is "near" to 0. Hence v_0 and $v_0 + v_1$ have different signs, and so v_0 and v_1 will have different signs.

If we solve the equation $\text{mlp}(x) = 0.5$ for x, we get

$$x = -\frac{\ln\left[-\dfrac{v_0 + v_1}{v_0}\right] + \omega_0}{\omega_1} = x^* \text{ say;}$$

x^* is the decision boundary where an observation is given an equal probability of being in either class.

We then find the maximum point of $\text{IC}(x, v_0)$ by differentiating with respect to x and setting the resulting expression to 0.

$$\frac{\partial}{\partial x} \text{IC}(v_0) = \frac{\partial}{\partial x}\left[\{\text{mlp}(x) - t\}\frac{\partial}{\partial v_0}\text{mlp}(x, w)\right]$$

$$= \frac{v_1 \omega_1 \exp(-\omega_0 - \omega_1 x)\text{mlp}^2(x)}{\{1 + \text{mlp}(x)\}^4\{1 + \exp(-\omega_0 - \omega_1 x)\}^2}$$

$$+ 2\frac{[\{1 + \text{mlp}(x)\}^{-1} - t]\text{mlp}^2(x)v_1\omega_1 \exp(-\omega_0 - \omega_1 x)}{\{1 + \text{mlp}(x)\}^3\{1 + \exp(-\omega_0 - w4x)\}^2}$$

$$- \frac{[\{1 + \text{mlp}(x)\}^{-1} - t]v_1\omega_1 \exp(-\omega_0 - \omega_1 x)\text{mlp}(x)}{\{1 + \text{mlp}(x)\}^2\{1 + \exp(-\omega_0 - w4x)\}^2}.$$

Setting this expression equal to 0 and solving for x, we get

$$x = -\frac{\ln\left[-\dfrac{K + v_0 + v_1}{K + v_0}\right] + \omega_0}{\omega_1},$$

where

$$K = \ln\left[\frac{1 \pm (1 - t + t^2)}{t}\right],$$

which in the case where $t = 1$ simplifies to

$$x = -\frac{\ln\left[-\dfrac{\ln(2) + v_0 + v_1}{\ln(2) + v_0}\right] + \omega_0}{\omega_1} = x^{**} \text{ say.}$$

Now if $v_1 \geq 0$ and $v_0 \leq 0$, then the curve $\mathrm{mlp}(x)$ is of the shape shown in Figure 8.3 and in order for $\psi_{v_0}(x)$ to be at a maximum when x is misclassified, we require that x^{**} be less than x^*; that is, we require

$$\frac{ln(2) + v_0 + v_1}{ln(2) + v_0} < \frac{v_0 + v_1}{v_0}.$$

Now as $v_1 \geq 0$ and $v_0 \leq 0$, this inequality can be easily shown to be satisfied.

In the case where $v_1 \leq 0$ and $v_0 \geq 0$, the fitted model will have a different shape (approaching "1" as $x \to \infty$ and "0" as $x \to -\infty$ and the inequality will be around the other way). Either way, the relevant inequality is satisfied, so that the maximum influence occurs when an observation is misclassified.

In the case of target values other than 1, the expression is different only in the constants, except for a target value of 0, which gives a somewhat different expression. However, the argument still reaches the same conclusion.

TOPIC INDEX